Journey Through

Air and Space

To the crew of 51-L who reached out and touched the face of God

Jane Hodges-Caballero, Ph.D.

Illustrations & Cover Design by Mauro Magellan

Edisto Press

Edisto Island, South Carolina

Edisto Press
P.O. Box 4
Edisto Island, SC 29438

This publication was made possible in part by a grant from the
American Association of University Women (AAUW)
Educational Foundation (Atlanta Branch)

ISBN 0-9618589-0-7
LIBRARY OF CONGRESS CATALOG CARD NUMBER: 87-81114

Printed In The United States Of America

Contents

Preface

JOURNEY THROUGH AIR AND SPACE represents a logical and well thought out approach for introducing space to youngsters. The importance of starting with a basic level, the earth and its characteristics, the atmosphere and what it provides and how man adapts to this normal environment is recognized. Then the youngsters can move logically from the friendly atmosphere into a space that we can only imagine and our total knowledge is bounded by what we hear, see or feel. We are totally dependent on books, movies, and word of mouth. This approach makes it real. I had the feeling of actually being there when I finished reading it. And even though I've been there, I actually learned a lot!

Colonel Frederick Gregory
Space Shuttle Pilot
NASA Astronaut

Acknowledgements

I would like to extend my sincere appreciation to the following persons and organizations for their continuing support and encouragement in promoting aviation and space education:

Dr. Alma David, University of Miami, Early Childhood Professor (retired)
Dr. Betty Rowen, University of Miami, Early Childhood Professor (retired)
Dr. Barbara Stevens, Director of Early Childhood/Middle Grades, Georgia Department of Education
Dr. Eleanor Wynne, President, Aerospace Education Foundation, Professor, Univ. of California (Irvine)
Ms. Barbara Jean Stevens, Kindergarten Teacher, Georgia Wing Civil Air Patrol, Aerospace Education Member
Mr. Kendrick Flint, Aerospace Engineer
Dr. Glenda Roberson, University of Texas, Early Childhood Professor
Dr. Anne Christenberry, Augusta College, Early Childhood Professor
Ms. Linda Hardie, Early Childhood Consultant/Instructor
Ms. Anne Williams, Director, Piedmont Academy North
Dr. Peggy Baty, Embry Riddle Aeronautical University Professor

USAF — Civil Air Patrol (USAF-CAP)
National Aeronautics and Space Administration (NASA)
Federal Aviation Administration (FAA)
Ninety-Nines (99s)
University of Aviation Association (UAA)
World Aerospace Education Organization (WAEO)
Young Astronaut Program (YAP)
Aerospace Education Association (AEA)

Original poems by Jane Caballero reprinted with permission from *Aerospace Projects for Young Children* and *Aviation and Space Folder Games for the Classroom,* Humanics Limited, P.O. Box 7447, Atlanta, GA 30309.

Introduction

This book is designed as a resource for teachers and parents concerned with early childhood education. It is hoped that the ideas and activities presented herein will serve in the creation of a space science and mathematics curriculum that is both child-centered and exciting. The basic philosophy for such a curriculum is based on that of Piaget:

> "The principle goal of education is to create men who are capable of doing new things, not simply of repeating what other generations have done — men who are creative, inventive and discovers. The second goal of education is to form minds which can be critical, can verify, and will not accept everything that they are offered. The great danger today is of slogans, collective opinions, ready made trends of thought. We have to be able to resist individually, to criticize, to distinguish between what is proven and what is not. So we need pupils who are active, who learn early to find out by themselves, partly through their own spontaneous activity and partly through materials we set up for them; who learn early to tell what is verifiable and what is simply the first idea to come to them." (Piaget, 1954)

As we begin our journey through space we need to understand what we are going to find, just like a pilot goes to ground school before he starts to fly an airplane. We need to have some information about our earth (physics) and environment (ecology) before we enter into this new environment of space (astronomy). (See SCIENCE section)

There is no weather in space. Why? What are clouds? How are they formed? This is what begins our space adventure. Concepts such as these will be discussed prior to entering space. What is in space? Planets and stars! How do we get into space? In rockets and Space Shuttles. The effects of the atmosphere on different flying objects is crucial if we are to learn about how to travel in space. Different objects fly differently depending on how the air affects the object. We will see how air affects kites, balloons, helicopters and airplanes. Then we can see how the Space Shuttle was designed to accommodate all these concepts.

A basic understanding of space must include things in space: clouds, stars and planets and things that fly: balloons, kites, airplanes, rockets and the Space Shuttle. Preparing to be an astronaut includes learning about what to wear, what to eat and how to keep fit. All activities will focus on the aspect of space adventure and emphasize sensory experiences to re-emphasize Piaget's belief that "Children must learn through the materials we set up for them": Sight (recognize properties, compare shapes, colors, puzzles, collages), hearing (sound, music, describing sounds); taste (sweet, bitter, salty and sour through cooking); and touch (contact with objects — warm, cold, pain, pressure, tracing, textures).

The aerospace curriculum will be initiated in the classroom by reading the first part of this book which establishes the basic understanding of math and science curriculum through sensory experiences.

The curriculum activities are based on the following format:

Name

Subject/Sense (Math or Science and which Senses are emphasized)

Skill

Procedure (The activity will be described with a list of required materials or resources.) If more than one activity is presented, choose one or more depending on your interests and the level of your children. Vary and adapt the activities as you wish!

Parent/Child Experience (An activity the parent or relative or friend can do at home with the child is described.)

Objective (What the child will do at home is listed.)

Background Information (Information is provided for the teacher)

File folder activities will help you become organized and prepare for the early childhood aerospace curriculum. These practical folders not only are inexpensive to develop but are easy to store and use in a learning center.

Making file folder games out of simple file folders, patterns in the book and colored marking pens are all you need to make your own activities. You can make variations as suggested or use your own imagination to individualize folders based on the students development level. These unique folders encourage the teacher/parent to take an active role in designing curriculum for the child. It is not a ready-made kit or book that merely requires presentation. Many other activities are provided to extend the folder concepts to the home and outdoor environment. These related activities will involve additional teacher/parent preparations and additional materials, i.e., bringing in a hotplate and teakettle to watch "clouds" form.

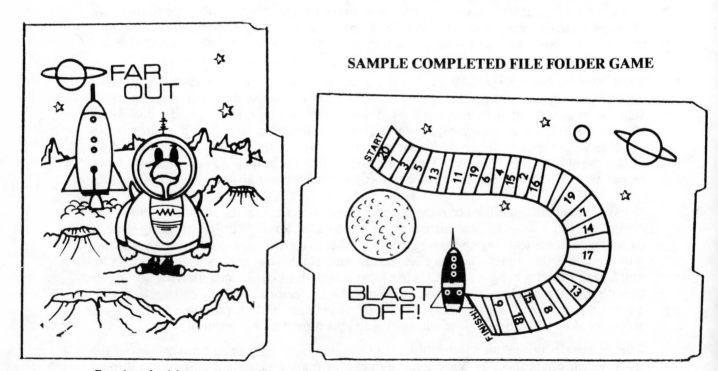

Reprinted with permission from Aviation and Space Folder Games, Humanics Limited

Piaget, Jean, *The Construction of Reality in the Child,* New York: Basic Books, Inc., 1954.

Math

MATHEMATICS IN EARLY CHILDHOOD

Early childhood educators know that young children learn by manipulating objects. However, it is difficult to tell someone what the child has learned.

Piaget identified two types of knowledge children learn when they manipulate objects. In science, it is known as physical knowledge (knowledge of objects that can be observed). In math, it is known as logicomathematical knowledge. Logicomathematical knowledge is created when we make relationships between objects, i.e. when we compare 2 balloons (1 red, 1 blue). Relationships between 2 objects are observable, but the relationship (same-different, number (two), color) are created in the person's mind. When the child builds with blocks, he looks for blocks to make his rocket taller. Thus, the logicomathematical knowledge is dependent on his physical knowledge.

Children learn through their senses. Piaget extended this belief by showing children can only obtain sensory information if they act on the object mentally and physically. The child must handle the object-manipulation is essential. The child will gain physical knowledge after interpreting with his logicomathematical framework the sensory experience.

We want children to have the opportunity to think. We need to create situations that are meaningful to the child, allow them to make decisions and give them the opportunity to talk with their peers to solve problems and exchange ideas.

We have tried to create choices in this book for the child and the adult. Basic skills and concepts in math and science will be discussed, but remember to create the activities with your students or children so sensory learning takes place.

Mathematical experiences in the early years should be based on a teacher knowledge of what to teach and understanding of the children to be taught. Piaget's work has helped produce programs that emphasize the development of broad intellectual powers, rather than the mastery of limited specific skills. He has identified certain stages of development. The preoperational stage of development will be the focus of the age group of this book. The preoperational stage begins about age one or two and lasts until approximately age seven. Learning must occur through play at this age. This is the stage of intuitive thinking. The child has difficulty with conservation. He needs many experiences to test and find out for himself. He needs to manipulate, explore and make choices. The real concrete experiences help him develop the thinking processes.

In the preoperational phase the child will classify things by a single feature (such as size). There is no concern about apparent contradictions (relationship between size and weight such as when a large light object floats and a small heavy object sinks.) The characteristic phrase is irreverability. The approximate age range is 2 to 4 years. In the intuitive phase intuitive classification emerges. There is gradual awareness of mass, weight and volume. (Amount remains the same even if transferred to a container of different size and shape.) The approximate age range is 4 to 7 years.

What should be involved in a math program for young children? The teacher needs to give attention to the development of number understanding. This is based on two levels of knowledge on the part of the learner:

Quantitative attributes and relationships of objects in the world (i.e., He has more airplanes than me. My rocket is higher than yours.)

Symbols used to represent these attributes and relationships (i.e., He has 4 airplanes and I have 2 airplanes.) To develop number understanding the classroom should provide the children with many concrete experiences to understand quantity and relationships before they are introduced to number symbols. Then number symbols should be introduced. Number understanding experiences in this book could be divided into the following goals:

Number and numerations: The ability to identify the properties of a collection which answers the questions how many or which one.

Measurement: The ability to express the size of a particular object.

Geometry: The ability to understand position in space. These goals could be broken down into the following concepts.

Number and numerations:
Sorts objects by similarities
Recognize a pattern
Arranges (orders) objects according to size, shape, color (seriation)
Matches objects of sets one-to-one
Recognizes/names number of items in a set up to five
Counts number of items in a set up to 10
Combines and separates sets of objects by a given characteristic (classify)
Selects numeral that names the number of elements of a set up to 10

Measurement
Compares two objects according to size
Measures length by counting nonstandard units

Geometry
Moves oneself, or an object from one point to another point using given directions
Sorts and identifies basic geometric shapes (circle, square, triangle, rectangle)

References:

Beihler, Robert F. *Psychology Applied to Teaching,* Boston: Houghton, Mifflin, Co., Inc., 1971

Georgia Department of Education, *Basic Curriculum Content,* Atlanta, Georgia, 1985

Georgia Department of Education, *Kindergarten in Georgia,* Atlanta, Georgia, 1978

Kamii, Constance, "How do Children Learn by Handling Objects?" Williams, Connie K. *Young Children,* Nov. 1986

Leeper, Sarah H., Skipper, Dora K., Witherspoon, Ralph L., *Good Schools for Young Children,* Macmillan Publishing Co., Inc., New York, 1979

Science

SCIENCE IN EARLY CHILDHOOD

Children are natural scientists with their spontaneous curiosity. They explore, discover and experiment with their environment. They need opportunities to develop skills in observing, questioning and solving problems. Children in the early childhood stage have already begun to explore their environment through their senses, but they don't have the thinking skills developed to understand their surroundings. Children will need to explore and manipulate objects in their environment to arouse their curiosity and interests. Then the adult can extend this curiosity to develop the child's language, knowledge and thinking abilities. Activities in this book will provide opportunities in:

Observing (find differences in objects, i.e. color, size, shape, texture)

Classifying (finds likenesses and differences, i.e. These fly. These do not fly)

Predicting (make predictions, i.e. These dark clouds mean its going to rain)

Reporting (describe observations verbally, i.e. Tang tastes different from fresh orange juice)

These goals can be broken down into the following concepts:

Sorts by shape, color, size, texture
Identified colors
Identifies objects by characteristics such as lighter, heavier-larger, smaller-floatable, non-floatable
Describes differences in weather from day to day
Identifies hot, cold-wet, dry-light-dark
Identifies sounds such as loud; soft-high-low
Identifies time frames such as day, night-yesterday, today, tomorrow
Identifies motion such as push, pull
Identifies and experiences activities using the five senses

Teachers will need to plan activities in the classroom that respond to the child's curiosity and interests as he is motivated through space science and man's relationship to this new environment. Space science involves experiences in physics (matter and energy), astronomy (our earth and the universe) and ecology (humans and our environment). Examples of concepts and activities in these areas follow.

Matter and energy (physics)
    Air is around us
    Air fills space
    Wind moves many things
    Heat changes some things
    We breathe air
    Magnets pull some things
    Sound travels
    Water evaporates and
      can change forms

Blow up balloons
Observe things moved by wind
Cook different foods
    observing changes
Listen to and identify sounds

Our earth and universe (astronomy)
- The moon, sun, and other stars are in the sky
- Man has put satellites in orbit around the earth
- Man has traveled into outer space
- Stars move in the sky
- Sun gives light and heat
- Earth is composed of water and soil
- There are different kinds of soil

- Collect different soil samples
- College pictures of where heat comes from
- Collect pictures of how water is used in the home
- Observe and measure shadows at different times of the day
- Compare objects left in the sun and those in the shade

Humans and our environment (ecology)
- Humans use plants and animals for food and clothing
- We use and control light
- We use and control heat

- Prepare foods obtained from animals/those obtained from plants
- Collect pictures of how we use light in the home
- College pictures of how we use heat in the home

Be responsive to the child's interests as you create this new curriculum. As you provide activities on space, respond to the child's questions. At lunchtime, have the children identify the types of food that they are eating, whether or not it could be eaten in space. During outdoor play, have the children observe the types of plants, insects, rocks and leaves that are outside and whether they could be found on the moon or another planet. They can also observe the weather conditions. The science of space differs from the science on earth and the children need to begin to see those differences as mentioned. Continue providing experiences to encourage them to question, "How is living in space different from living on earth?" Consider the following three areas in encouraging these experiences.

The environment must be taken with you. There is no food, water, air or temperature in space. Spatial relationships change as we live in our new environment. Disorientation occurs in space. Up and down don't exist because of the absence of gravity. Weightlessness is a concept that will be difficult to understand, but awareness of the concept of gravity and how it affects our life will be emphasized through activities in this book.

As your study of aerospace begins, the following definition may be a helpful reminder of this field of education:

"Aerospace education is that branch of general education concerned with communicating knowledge, skills, and attitudes about aerospace activities and the total impact of air and space vehicles upon society."
Dr. Mervin K. Strickler

References:

Georgia Department of Education, *Basic Curriculum Content,* Atlanta, Georgia, 1985

Georgia Department of Education, *Kindergarten in Georgia,* Atlanta, Georgia, 1978

Leeper, Sarah H., Skipper, Dora K., Witherspoon, Ralph L., *Good Schools for Young Children,* Macmillan Publishing Co., Inc., New York, 1979

Senses

SENSORY LEARNING EARLY CHILDHOOD

Experiences in aerospace should be integrated throughout all developmental areas: physical, social, emotional and intellectual (math and science are emphasized in this book). Children feel good about learning space concepts. The activities in this book will get the teacher started in creating an exciting environment for the child to explore physically and mentally through his senses. Activities can be individual or group, depending on the development and interest of the child. Social skills will be enhanced as the children play the activities together. Cognitive development is based on language development. Experiences such as field trips, dictating language experience stories, seeing charts and pictures, participating in dramatic play and drawing and copying are suggested to enhance cognitive development. All these experiences are learned through the child's senses. Activities in this book are designed as sensory experiences and the following information on the senses is provided for background information.

SIGHT
The eye is our most important organ for finding out about the world around us. We use our eyes to enjoy things around us and help us carry on daily activities. We see the different properties of objects and learn to classify them, compare them, and group them, thus helping us to understand our world better. The eyes are like cameras and can set themselves automatically for light, speed and distance. The eye is made up of three layers. The retina is the inner layer of the eye and is sensitive to light and color. The cornea is the front part of the eye's two outer layers. It is clear and permits light to enter the eye. The pupil is the opening in the middle layer — it is the black spot. The iris is the thin curtain of tissue which is in the front of the lens. Eyes also have eyebrows over them and eye lashes around them. The eyebrows and eyelashes are made of tiny hairs that keep dust from falling into the eye. Eyes have eyelids that close over the eyes to keep out the light and dust. Very young children can have their eyes checked with an E chart (which shows the directions in which the letter is pointing) to see whether or not glasses are needed.

HEARING
The ear is a delicate organ that can be easily damaged by infection or injury. We should be careful not to put anything in the ear or we could become partly or completely deaf. Our ears help us learn to talk. We hear high and low, loud and soft sounds. Sounds are caused by something moving back and forth very fast called vibration. A vibration uses sound waves in the air. We cannot see sound waves but we can see what causes them. When wind blows over the leaves, we can hear a sound. The eardrum in each ear starts to vibrate when a sound wave hits it. The eardrum causes three little bones to vibrate. The nerves in our ears pick up the vibration and carry the messages to the brain. The parts of the ear are: eardrum, inner ear, middle ear and outer ear. The three small bones are found in the eardrum and called the hammer, anvil and stirrup.

SMELL

Smell is one of the most important senses in man or animal. Like sight and hearing, smell gives us information about our environment. The nose must come into contact with an odor. The olfactory cells are the tiny hairs on the upper part of our nose. Nostrils are the two little openings at the end of the nose. Air goes up the nose and into the head. When it gets into the head, you can smell it. There are three basic types of odors. They are difficult to identify and organize. They are: flowery, fruity and burnt odors. Smell helps us distinguish one food from another and taste the food.

TASTE

Our tongue helps us tell one food from another. The taste buds are bumps on the tongue. They include sweet taste buds on the front of the tongue, bitter on the back of the tongue and salty and sour taste buds on the sides of the tongue. We learn to organize foods and drinks by the way they taste. Sweets, i.e. candy, cakes, ice cream, are identified by the front of the tongue. Salty items, i.e. pretzels, are identified by the front side of the tongue. Sour items, i.e. lemons, are identified by the back side of the tongue. Look at your tongue. Can you see that the taste buds on the front of the tongue are smaller than those at the back of the tongue?

TOUCH

Touch is the sense which gives us the sensation that we have come into contact with an object. We learn the shape, and softness or hardness of an object by touching it. Touch is pain, pressure, smooth, soft and hard. The senses are in the muscles and bones. We can feel with any part of our bodies. We can find out about objects by feeling similarities. Different parts of our bodies are more sensitive to touch than other parts. We can organize things by the way they feel. The feeling of pressure has the biggest number of sense organs. It is more developed on the tip of the tongue and least developed on the back of the shoulders. Finger tips and the tip of the nose are also sensitive areas. Nerve endings form small discs, just inside the living layer of the skin. These nerve endings are around each of the hairs on the skin.

References:

Caballero, Jane *The Handbook of Learning Activities for Young Children,* Humanics Limited, Atlanta, Georgia, 1980

"Integrated Components of Appropriate and Inappropriate Practices for Four and Five Year Old Children," *Young Children,* September, 1986

CLOUD MATCHING

NAME: Clouds

SUBJECT/SENSE: Science (Sight, touch)

SKILL: Become aware of various cloud forms

PROCEDURE:

1. Clouds are vaporous formations in the sky. They can be made in the following way: Put water in the teakettle, put teakettle on hotplate or stove, let the water in the teakettle come to a boil. Watch the escaping vapor condense as it rises and cools, forming a small cloud. Share the observation encouraging the children to tell what happened.
   Materials: stove or hotplate, teakettle, water

2. Pour two to three inches of hot water in a jar. Cover jar with the lid. Put three to four ice cubes on the lid and watch the clouds form. Share the observation encouraging the children to tell what happened.
   Materials: jar, water, ice cubes

3. Provide the children with cotton balls, blue construction paper and glue stick. Encourage the children to tear the cotton apart and create cumulous (puffy), stratus (layers) and/or cirrus (curly, featherly) cloud formations, then glue the formations to the paper.
   Materials: cotton balls, blue paper, glue stick

4. The teacher can beat a mixture of 2 cups Ivory soap powder and 1¼cups water. The children can fingerpaint and form clouds on blue construction paper.
   Materials: Ivory soap, water, bowl, beater (fork), blue construction paper

5. The teacher will create the folder game activity on the following page. Attach a brad next to all cloud formations. Wrap a string around the brad on the left column. Allow the children to match the clouds by wrapping the string around the brad on the right column.
   Materials: folder, pattern, glue, brads, string

6. A large pool of water is below the Space Shuttle on the launch pad to help cool the rocket engines at blast-off. The Space Shuttle makes a great big cloud when it blasts off. The heat from the engines heat up the water around the Shuttle creating a giant cloud. Look at NASA photos of the Space Shuttle launch so the children can see the big clouds.
   Materials: photograph of Space Shuttle launch

PARENT/CHILD EXPERIENCE:

Go outside and look at the clouds with your child. Ask your child what he thinks the clouds mean. If some clouds look darker than others ask your child if he knows why some clouds look darker than others. If there is a dark or thicker cloud (it is more dense allowing less sunlight to pass through) it probably means it is going to rain because it has more moisture in it. Ask your child to talk about how the rain sounds.

OBJECTIVE: Observe various clouds

BACKGROUND INFORMATION:

There are no clouds in space; however, it is very important to learn about clouds if you are going to travel in air or space because a pilot can't tell if he is up or down when he flies into a cloud. Cumulous clouds are near the earth's surface or can tower thousands of feet in the sky. Stratus clouds form in layers like fog and are seldom above 6,500 feet. Cirrus clouds are in the top level of the atmosphere (troposphere) which is below 50,000 feet. Space is above the earth's atmosphere, (troposphere) where there is no moisture or air that is needed to form clouds.

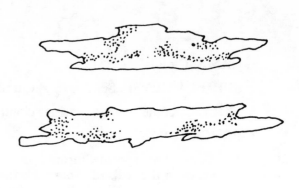

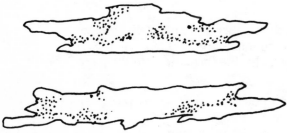

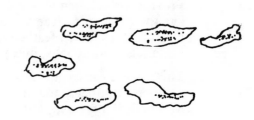

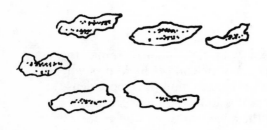

16

17

NAME: Weather

SUBJECT/SENSE: Science (Sight, touch)

SKILL: Become aware of weather (rain and air) and that weather changes

PROCEDURE:

1. The teacher will make the folder game on the following page. Discuss the seasons with the children and that certain symbols represent certain types of weather — snowman with snow represents cold weather; umbrella represents rainy weather, a kite represents windy weather; and the sun represents hot weather. Then ask the children to match or say the word that represents the weather symbol.
   Materials: Folder, pattern, glue, markers, scissors

2. To show how air will cool as it expands outward and upward, ask the children to put their hand close to their mouth and blow out. Then ask them to hold their hand about eight inches away from their mouth and blow out (exhale). Ask the children if they could tell a difference in the air as it struck their hand at different distances from their mouth.
   Materials: None

3. Pour water into two different glasses. Place a thermometer in each glass. Place one glass in a shady spot and one glass in direct sunlight. Have children observe the mercury in the thermometer rise to a higher temperature in the sunlight than the one in a shady spot. This will demonstrate how sunlight gives us heat. They can also put their finger in the water to feel the differences in temperature.
   Materials: two glasses, two thermometers, water

4. Discuss the weather equations with advanced children. They will learn the symbols that represent sun, air, moisture (rain), rotation.
   Materials: folder, pattern, glue, markers, scissors

5. To find out how much it rained during a downpour, the teacher will help children to mark one-half inch lines along a piece of masking tape using a ruler. Attach masking tape vertically along a jar. Place jar outside in an open space. Look at the jar and measure how much it rained.
   Materials: jar, masking tape, market (pen), ruler

6. The teacher will light a piece of paper and drop into a glass bottle (the neck should be a little smaller than an egg). Put a moist-shelled hard boiled egg on top of the glass bottle. The egg will fall into the bottle because air expands when hot and contracts when cool. To remove the egg have child blow into inverted bottle, the egg will fall out due to change in air pressure.
   Materials: hard boiled egg, glass bottle, paper, match

PARENT/CHILD EXPERIENCE:
Hot air expands. This can be demonstrated by the following experiment. Blow up a balloon. Tie it with a string. Tie it to a hot radiator or near a sunny window. When the balloon gets hot, it will burst because hot air expands. Ask your child to tell what happened.
Materials: Balloon, string

OBJECTIVE: Observe hot air expanding

BACKGROUND INFORMATION:
Space weather satellites have found out that there is no weather in space because there isn't any air or water. Weather satellites have provided tremendous knowledge about the current and predicted weather. Millions of lives have been saved because of more accurate weather predictions. It is very important to understand weather before blast-off into space. Spacecraft have to blast-off in good weather and it must be predicted that the weather will be good upon reentry. Depending on weather conditions the Shuttle will land at Kennedy Space Center in Florida or Edwards Air Force Base in California. There are also alternate landing spots on the other side of the earth if unexpected weather occurs. Discuss with your children how they would feel about living in space without weather changes.

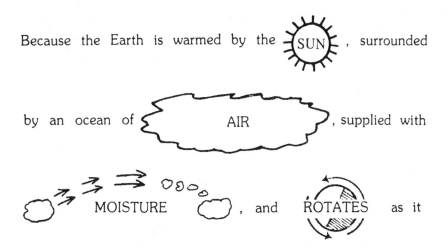

travels in space, we have many changes in our weather. Changes in the weather are important to pilots because they want to operate their aircraft safely. Many times the Space Shuttle pilot did not take off or land because of bad weather.

THINK ABOUT THE WEATHER

1. Is there weather on the moon?

2. Why or why not?

3. Why is knowledge of the weather important to a pilot?

4. Complete this equation:

Print the following words on the strips of construction paper and match with the appropriate weather symbol: WINDY, RAINY, COLD (SNOWY), HOT (WARM)

ANSWERS

1. No.

2. The moon has no air or water.

3. The pilot wants to operate his plane safely and efficiently

4. A. Sun, B. Air, C. Water, D. Rotation of Earth.

20

CONSTELLATIONS

NAME: Constellations

SUBJECT/SENSE: Science (Sight), Math (touch)

SKILL: Identify stars and that stars give off light; count to seven and learn that stars give off light

PROCEDURE:

1. The teacher will read the story about the Big Dipper then tell the children about the Big Dipper. Ask the children to count how many stars are in the Big Dipper. Allow the children to use stick-on stars to paste on top of the formation of the Big Dipper (one-to-one correspondence).
Materials: folder, pattern, glue, stick-on stars

2. The sun is our closest star. Color comes from light and sunlight is made of many colors. Show that sunlight is made of all colors of the spectrum using a prism. The prism bends the light and divides the colors into separate sections. Children can name the colors.
Materials: prism (crystals)

3. Turn on a flashlight in a darkened room and outside in the bright sunlight. Ask the children to tell which light is brighter. This will illustrate to the children that stars are seen only at night because the sunlight is so bright during the day that you cannot see the stars.
Materials: flashlight

4. The teacher can cover a paper towel or toilet paper tube, including one end with tin foil. Punch holes in the end with a small pointed object. The child can hold the toy telescope up to the light seeing the 'stars' and singing 'Twinkle, Twinkle, Little Star' or 'Star Light, Star Bright.'
Materials: paper towel, toilet paper tube, tin foil, any pointed object

PARENT/CHILD EXPERIENCE:

Go outside and observe the constellations with your child on a clear night. See if you can locate the Big Dipper.

OBJECTIVE: Observe stars in the sky

BACKGROUND INFORMATION:

Stars can be seen better in space and they do not twinkle because you don't have to look through the earth's atmosphere. (In fact, you can see 25 times better!) Stars help the Space Shuttle pilot know his position. He can look at the North Star and relate to his position on earth. Stars actually created their light years ago because light travels at 186,000 miles per second. The light that we presently see from the sun left the sun about 8 minutes ago.

THE BIG DIPPER

The most recognizable constellation in the Northern Hemisphere is the Big Dipper.
It is shaped like an old-fashioned water dipper. (The Big Dipper is part of the Great
Bear, a larger constellation). The Big Dipper contains seven stars. The middle stars
of the handle (Alcor and Mizar) appear to be a double star. The two outer stars of
the bowl point to the North Star (Polaris) Since a magnet does not work in space.
the North Star helps pilot-astronauts determine their position because it is always
in the north.

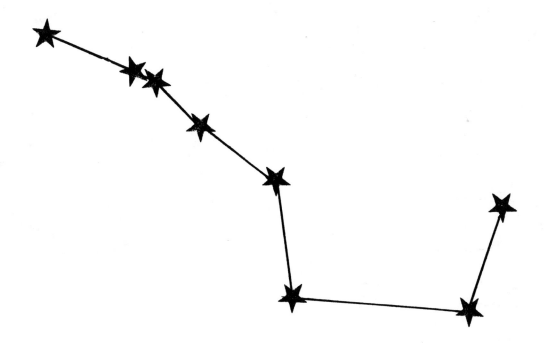

SOLAR SYSTEM

NAME: Solar System

SUBJECT/SENSE: Science (Sight)

SKILL: Become aware of the solar system (there are planets and a sun in the sky and that we live on one of the planets called earth); the sun gives us heat and light.

PROCEDURE:

1. The teacher will assist the children in making a solar system out of paper, balloons, paper mache, or styrofoam balls. The teacher will arrange the planets in correct orbital position as the children count the planets. Strings can be attached to "planets" and hung from a coat hanger.
    Materials: Nine balloons, nine paper mache balls or nine styrofoam balls, coat hanger, string (if paper mache or styrofoam is used children can use tempera paint, water and brushes to paint the planets.)

2. The teacher will create the folder game activity on the following page. She will color two sets of planets to match. She will laminate one set on file folder in correct orbital position. The teacher should number the planets on the back of the second set so the children can arrange the planets in number sequence. The children will match the color and size of the circles.
    Materials: folder, pattern, glue, scissors, markers

3. The teacher will read the poem to the children and then have them repeat the poem.

    Planets
    Mercury, Venus, Earth and Mars
    These are the planets that dwell near the stars.
    Jupiter, Saturn, Uranus, too, Neptune and Pluto,
    I know them, do you?

    Earth
    The earth is a great big ball (Make round circle with hands)
    It isn't flat at all (Hold palms together)
    It spins around like a top (spin finger)
    It never ever stops (shake forefinger)

4. Use a flashlight to represent the sun and styrofoam balls to represent the planets. The teacher will shine the flashlight as the children rotate the planets slowly on their axis (their hands) and orbit (walk around the sun). The children will watch how the sun gives the planets light.
    Materials: flashlight, styrofoam balls

PARENT/CHILD EXPERIENCE:
Go outside with your child at different times during the day. Watch the effects of the sun casting a shadow at different times of the day. This will show your child how the sun changes the amount of heat and light provided during the day.

OBJECTIVE: See that the sun gives us heat and light as it shines on earth

BACKGROUND INFORMATION:

The Space Shuttle is exploring our solar system. The sun is the center of our solar system and the planets revolve around the sun. Space probes (satellites) have visited other planets. Pictures of the planets can be obtained from NASA. (See Resources)

Mercury — closest to the sun; yellow in color; life as we know it isn't possible because there is no air or water and its too hot or too cold depending on which direction it is facing.

Venus — closest to the size of the earth; life isn't possible as we know it; we call it our morning or evening star

Earth — only planet that has life on it; one moon; length of each day is twenty-four hours

Mars — two Viking spacecraft landed on Mars on July 20, 1976, found rolling dunes of orange dust and volcanic rocks; water may have been on Mars about a billion years ago or may be underground; took many exciting photographs and performed many experiments

Jupiter — Pioneers 10 and 11 spacecraft visited Jupiter March 1972 and April 1973; discovered rings; largest planet with 12 moons

Saturn — Voyager spacecraft verified 9 moons and over 100 rings on a recent visit to Saturn

Uranus — The Uranus Encounter, the most recent planetary visit discovered ten additional satellites, bringing the total number of satellites (moons) to fifteen, verified ten rings around Uranus; green in color

Neptune — Voyager spacecraft is on its way to Neptune; has two moons that go in opposite directions

Pluto — farthest planet from the sun, however, there could be a planet even further according to scientists; one moon called Charon

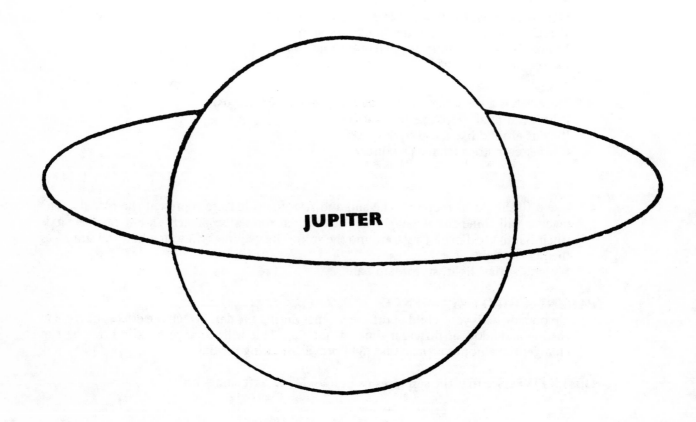

PLANETS

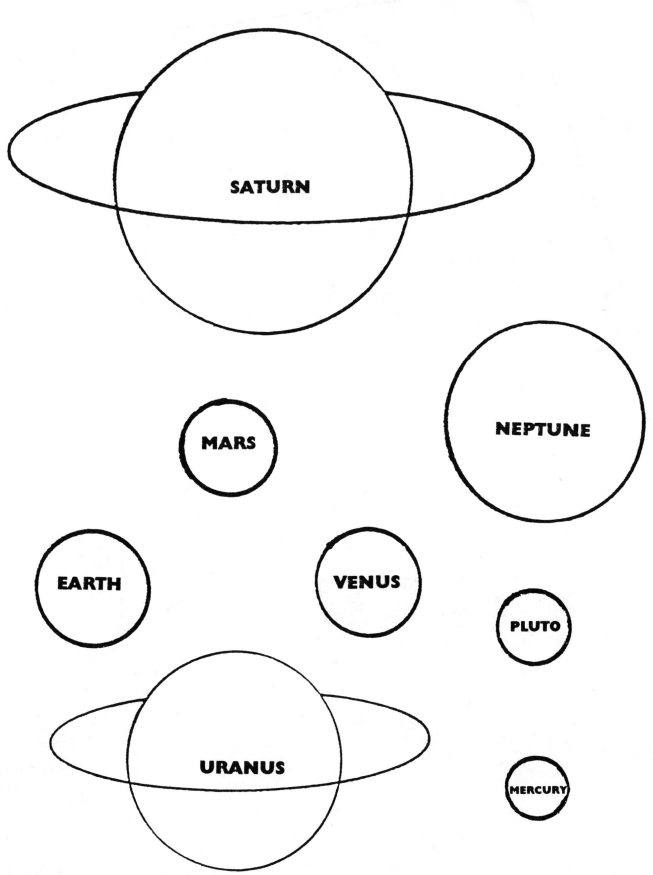

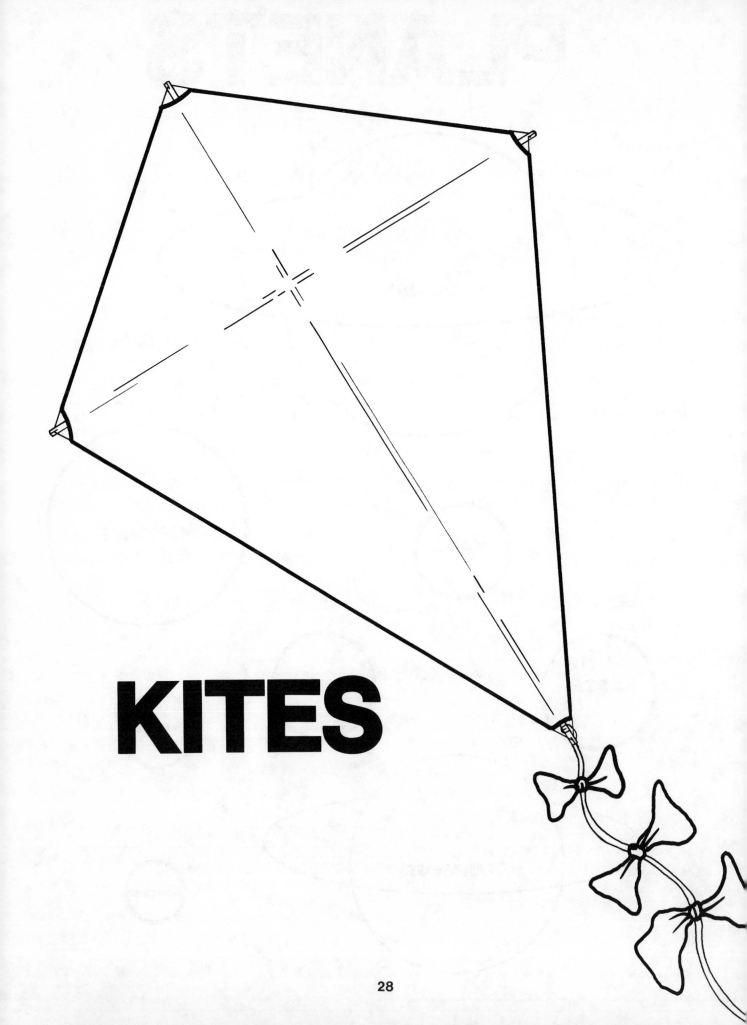

KITES

NAME: Kites

SUBJECT/SENSE: Science (Sight, touch)

SKILL: Become aware of the effects of moving air, wind, on an object; visually discriminate through matching

PROCEDURE:

1. The teacher will have the children make their own kites using two triangles (a diamond) and popsicle sticks. The children can display them on a bulletin board or hang them from the ceiling.
   Materials: two paper triangles, string or yarn, colors, markers, paints, some scrap material for a tail

2. The teacher will create the folder game activity on the following page. Attach a brad next to all kite formations. Wrap a string around the brad on the left column. Allow the children to match the kites by wrapping the string around the brad on the right column.
   Materials: folder, pattern, glue, brads, string

3. The teacher will make a sailboat with the children. Put clay or wax in a walnut shell. Glue or tape the triangle onto a toothpick and stick it in the clay or wax. Have the children place the walnut in a bowl of water and watch the effects of the boat as they blow on the triangle (sail) in different directions.
   Materials: triangle, tape or glue, toothpick, clay or wax, walnut, bowl of water

4. Talk about how kites are used by different people at different times. A long long time ago the Chinese built kites (1,000 B.C.) The Japanese then made kites for religious ceremonies. Stories were told of men being carried by kites. Experiments were done with manned kites (19th and 20th centuries). Today many people enjoy soaring through the air with a sophisticated kite (hang-gliding).

5. Make the pinwheel using the pattern. As the child blows on the pinwheel, the effects of air will be observed.
   Materials: pattern, scissors, pin, straw, tape

PARENT/HOME ACTIVITIES:

1. The parent will read the safety rules of kite flying to their child.

2. Build and fly a kite with your child. (see directions)

   Be sure and attach a tail to the kite. It keeps the bottom of the kite down providing stability. If you don't have a tail, the kite spins. The length of the tail can be adjusted to provide the stability needed.

OBJECTIVE: Fly a kite seeing the effects of air

BACKGROUND INFORMATION:
Children need to learn about the effects of air (airfoil) on various flying objects, i.e., kites, balloons, airplanes, helicopters. The Space Shuttle flies differently depending on where it is. It is like a rocket at blast-off, like a spaceship in space (wings have no effect in space), and like a glider with no engine when it returns to the earth's atmosphere. It is like an airplane because it has the flight controls of an airplane.

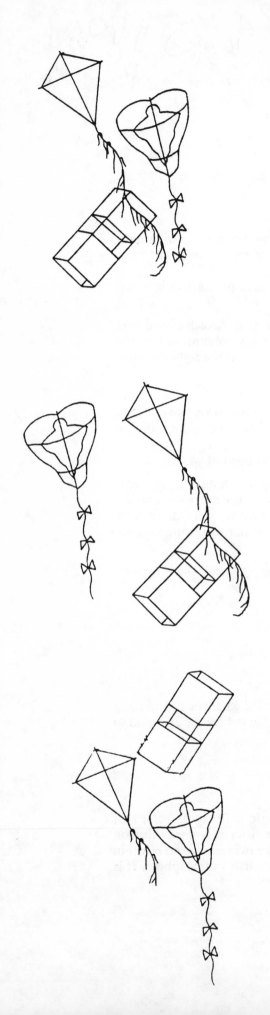
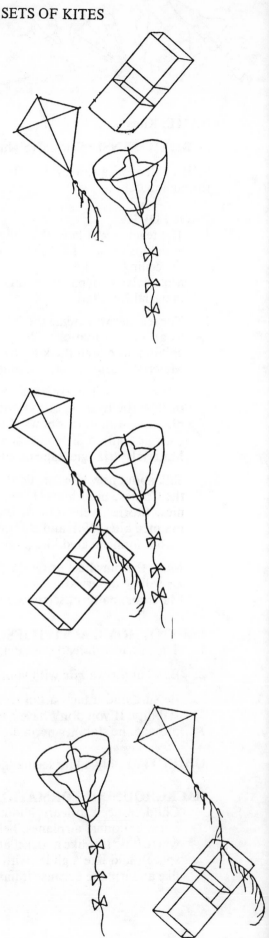

ACTIVITY

To make your own kite, follow the directions below. The following guidelines are provided for safe, fun kite flying:
1. Fly your kite far from power lines.
2. Always use dry string, wood and paper in your kite — never use metal, wire or metallic string or cloth.
3. Don't fly your kite in the rain.
4. Don't cross streets when kite flying.
5. Always fly your kite away from antennas.
6. Always call your power company if your kite gets snagged in a power line. Do not pull the string or climb power poles.

STICK KITE

Materials   2 sticks — one 36 inches long, the other 30 inches long.
Covering — newspaper, tissue, plastic, cloth, almost anything on hand
Kite string
Glue or tape

Procedure   Make saw cuts or ridges in the four ends of the sticks. Lash and then glue cross stick about 9 inches from the top of the longer stick.
Run string around the kite through each cut at the end of each stick and tie.
Lay the frame on the covering, leaving a margin of 2 inches all around. Glue the edges down over the string.
Bridle· The bridle should be 1-1½ times the length of the vertical stick. Attach one end of the bridle to the intersection of the two sticks and the other end to the bottom óf the vertical stick. Tie the flying line 4/5 of the distance up from the bottom of the kite.
Tail· Make a tail by attaching scraps of paper or cloth to a 10 inch string.

Reprinted from NASA Fact Sheet

PINWHEEL
Cut on solid lines.
Fold every other point toward the middle.
Stick a pin through the points and a soda straw.
Bend pin point and secure with tape.
Blow on the pin-wheel, seeing the effects of air.

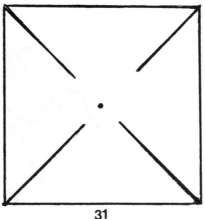

BALLOONS

NAME: Balloons

SUBJECT/SENSE: Science (Touch), Math (Sight)

SKILL: Become aware that air (wind) moves objects in the sky; create patterns

PROCEDURE:
1. The teacher can talk about the first hot air balloon to the children. Ask the children questions such as ''Who were the first living passengers?'' to increase listening comprehension.
   Materials: folder, story

2. The teacher can color the worksheet, creating repetitious color patterns, i.e., red, blue, red, blue, etc. Ask the children to name the colors. Then the children can color their own balloons with repetitious patterns. The teacher will color a second balloon the same color. Cut out the strips and have the children match the strips of color to the balloon.
   Materials: folder, pattern, glue, markers

3. The teacher can help the children cover a blown-up balloon with paper mache. Paint and decorate the balloon. Attach a basket at the bottom; hang from the ceiling. (See following page for directions.)
   Materials: paper mache, balloon, paint, decorations, basket, string

4. The teacher can buy a helium balloon from a mall or stove. Untie the knot and inhale a little bit of the helium — then talk. You will sound like Donald Duck. Air across the vocal cords causes sound. Helium is less dense so it causes a variation in sound. The children will hear the effects of air.
   Materials: helium balloon

PARENT/CHILD EXPERIENCE:
1. Have your child blow on a dandelion (spring or fall). Have your child watch the flight of the dandelion and compare it to balloons and birds in flight.
   Materials: dandelion, balloon

2. Help your child fill one balloon with air and another with water and tie the balloons. Have your child take the two balloons outside and drop them at the same time to compare which is heavier; lighter. Explain the concept of gravity to your child — heavier objects fall faster on earth.

OBJECTIVE: See the affects of air and gravity

BACKGROUND INFORMATION:
A balloon can rise because a gas is used inside the bag which is much lighter than the air around it. The earliest balloons were filled with heated air. Since hot air is less dense than cold air, the balloon would rise. There is a hole in the top of the balloon to let the hot air molecules escape making the air inside the balloon lighter. However, the air would soon cool so the ascension of hot air balloons is short. The Montgolfier bothers experimented with the first hot air balloon. Three months later they were invited to put on a demonstration for King Louis XVI and Marie Antoinette in France. They attached a cage to the balloon which contained a sheep, a rooster and a duck. The animals stayed aloft eight minutes and traveled a mile and a half. These animals became known as the first passengers in flight! Riding a balloon is a sport for many people who enjoy the thrills and adventures of floating in the air. There is much skill required in locating various currents of air and maintaining the proper altitude to fly with certain paths of air.

PAPER MACHE BALLOON

Materials

newspaper
wheat paste (consistency of whipping cream)
tempera paint and brushes
decorations (sequins, lace, ribbons, etc.)
pint milk carton
string

Procedure

Tear the newspaper into strips 1 inch x 6 inches
Blow up the balloon
Dip the strips into the wheat paste, wipping off excess
Cover the balloon with two layers of paper strips
Let the paper mache completely dry
Paint and decorate the balloon
Attach strings around the balloon
Attach ends of strings to milk carton
Hang balloon from the ceiling

During the time that the letter "B" presented, helium balloons with a simi message may be launched:

*Hello! My name is—————. If you fir my balloon, please write me at:
——————————(school/center)
————————(street)
————————(city,state)*

We have been studying the letter 'B' and launched balloons on Friday, Apr 13,1986. Thank you for writing to me.

Dear Isaac,
My name is Dana I live in Fa burn, Georgia. I go to kindergarten ju like you. I found your red balloon i my yard.

Your friend,
Dana

Contributed by Rita Melville, Newnan Maggie Brown Kindergarten

PARACHUTES

NAME: Parachute

SUBJECT/SENSE: Science (Touch)

SKILL: Become aware of drag (air resistance) on a falling object

PROCEDURE:
1. The teacher can make the parachute according to the directions, then have the children toss the parachute into the air, observing it as it falls. Children learn about drag as they watch the parachute fall. (Parachutes are also used on some race cars and some aircraft as they land to increase drag and help them stop.)
    Parachute Directions: Cut 8″ piece of thread. Wrap thread around 4 corners of paper towel square. Tie the thread ends together with a weight (paper clip).
    Materials: Kleenex or paper towel square, thread or string, paper clip (weight)

2. Talk about how parachutes are used in the space program. See background information.

PARENT/CHILD EXPERIENCE:
Find different scraps of cloth at home. With your child make parachutes out of different materials and add different weights. Have your child compare the difference as the parachutes fall. (The cloth will be too heavy unless a heavy weight is used.)
Materials: Scraps of cloth, string, weights

OBJECTIVE: Be able to drop the parachute seeing the effects of air on a falling object

BACKGROUND INFORMATION:
The Apollo capsule had to have a parachute as it re-entered the atmosphere to slow it down as it hit the water. The solid rocket boosters on the Space Shuttle have to have parachutes to slow them down as they are dropped into the ocean. Parachutes are also used to save your life if an aircraft fails. Packing a chute is very important and you have to have a license to do it. However, there is a reserve chute just in case the main chute fails. Astronauts cannot eject from the Space Shuttle and parachute out because of many reasons; economics, lack of space, too much weight, lack of time at blast-off, etc. Many people today parachute as a hobby with colorful sport parachutes. They participate in competitions based on style and accuracy.

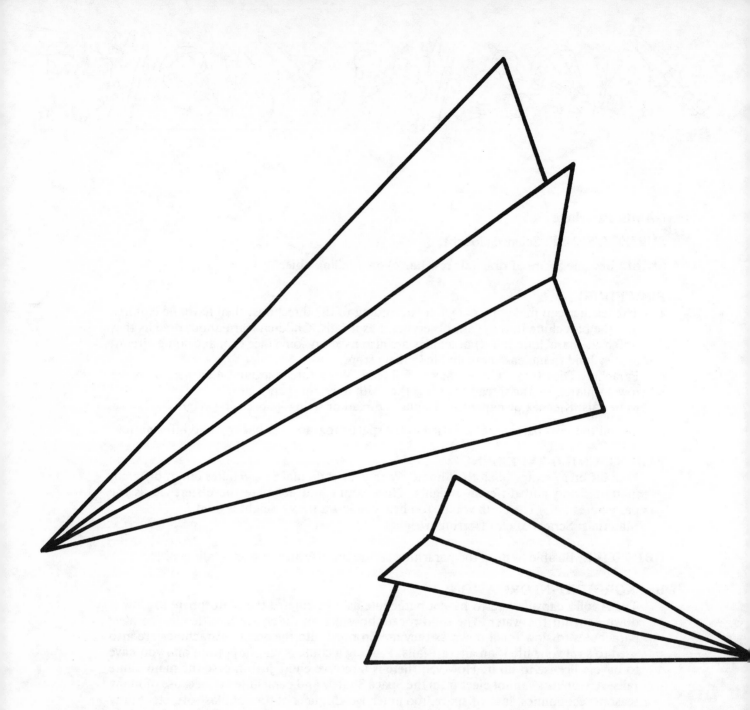

FLIGHT

NAME: Flight

SUBJECT/SENSE: Science (Touch, sight)

SKILL: Become aware of how aircraft fly, the effects of air on a moving object, the effect of air over an object (lift)

PROCEDURE:
1. The teacher can make the paper helicopter. Then the children can dip the helicopter to see how it flies. The helicopter spins as it floats downward.
   Materials: folder pattern, scissors, paper clips

2. The teacher can make the folded paper airplane and the children can fly it with the teachers directions. Have the children compare it to the paper helicopter flight
   Materials: sheet of plain standard size paper (8½ x 11)

3. The teacher will cut out a piece of paper and the children will blow over the top of it to see how it rises. This will illustrate lift to the children (see Background Notes).
   Materials: paper and scissors

4. With the children visit a local airport to see a helicopter and compare it to an airplane.

5. Trace the airplane body, wing and tail onto a file folder. Pull the wing through the slot on the airplane and the tail into the rudder slot. Place a paper clip on the nose. Have the children fly the plane and see what maneuvers they can make.
   Materials: folder game pattern, scissors, cardboard

PARENT/CHILD EXPERIENCE:
Help your child draw an airplane, noting the front (that thrust makes it move forward) and top (that lift makes it go upward). Talk about how an airplane flies. (An airplane flies because air lifts it up, gravity pulls it down and the propeller and engine pull it forward.)
Materials: paper, crayons

OBJECTIVE: Draw an airplane

BACKGROUND INFORMATION:
The important principle to understand as to why an airfoil can produce lift is to consider Bernoulli's law. He proved that where the speed of a moving gas is high, the pressure is low. Where the speed is low, the pressure is high.

Helicopters were used to find and pickup Gemini, Mercury, and Apollo capsules as they re-entered the earth's atmosphere and fell into the oceans. Helicopters are used to track the Shuttle and find the solid rocket boosters as they drop into the ocean.

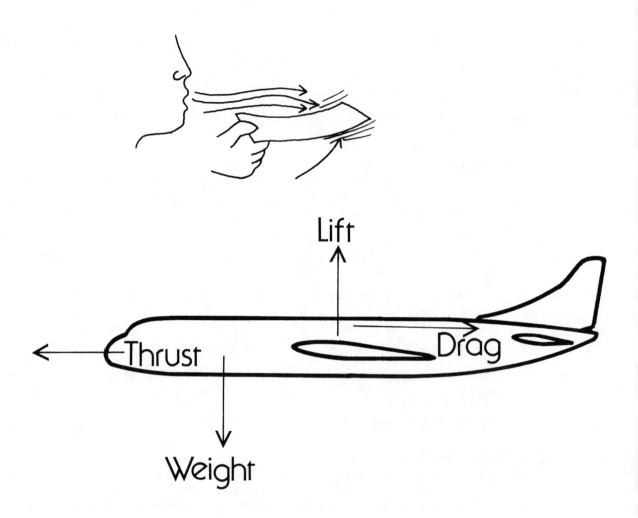

THEORY OF FLIGHT

A simple experiment will help you see how *Bernoulli's law,* works to produce lift.

Cut a piece of paper two inches wide and seven inches long. Hold it against the chin under your bottom lip with the narrow part. Then blow hard over the top of the paper. The paper rises!

What actually happens is the "air in a hurry" on top of the paper has less pressure. The pressure under the paper is greater and lifts the paper up.

HELICOPTER

Have you ever watched the aerobatics of fall leaves floating in the wind ?
Why do some leaves spin, some float and some glide?
Gather an assortment of fall leaves and compare their size, color, shape.
Compare other natural objects that blow in the wind, i.e. seeds.
Compare the leaves or seeds dropping in the air to the paper helicopers.

Cut on solid lines.
Bend on dotted lines and secure   with a paper clip.
Fold flaps at the top in opposite directions.
Drop the helicopter and watch it twirl.
The child can watch the effects of air on a falling object.
The helicopter will spin and spin.

Variation: Make 6 cuts down the sides of a paper cup
stopping one inch from the bottom.
Fold the strip outwards.
Flip the helicopter as you would a frisbee.

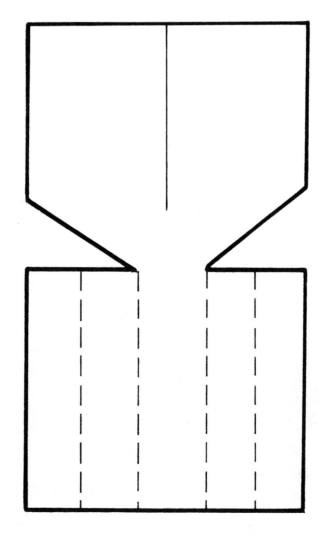

AIRPLANES

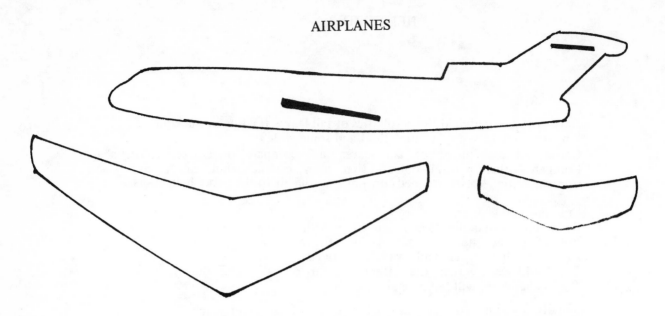

Variation: Allow the child to cut out his own airplane body and wing
designs out of styrofoam meat trays. (The teacher will need to
cut the insert in the body for the wing using a razor blade
or mat knife.) Use a paper clip on the nose to help stabalize
the plane.

Paper Airplane

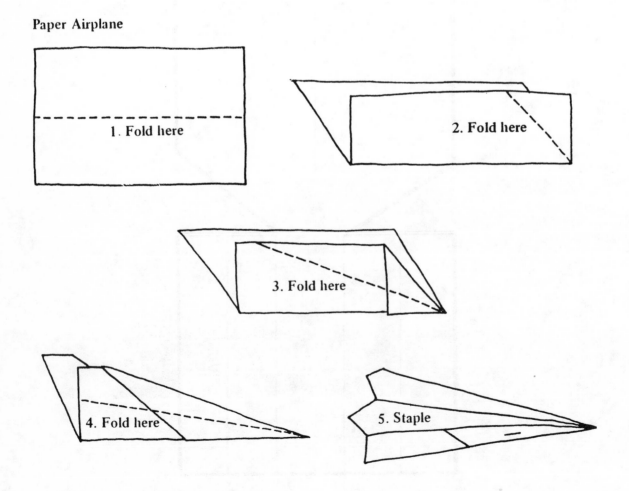

1. Fold here

2. Fold here

3. Fold here

4. Fold here

5. Staple

PILOTS

NAME: Pilots

SUBJECT/SENSE: Science (Touch, Hearing, Sight)

SKILL: Develop the ability to follow directions; enhance visual discrimination, become aware of being a pilot

PROCEDURE:

1. The teacher will explain to the children how a trip is planned on the ground using a road map. Pilots plan a trip in the air using an air-way map that shows roads and markers as seen from the air. The pilot files his flight plan so air traffic controllers can tell other pilots where he is flying to avoid an accident. Then if an accident occurs people will know where to look for him. (FAA investigates crashes and CAP performs most search and rescue missions.) Invite a pilot to your classroom to talk about a flight plan.
    Materials: Abbreviated flight plan (See next page)

2. A pilot must fly in a traffic pattern whether flying in a small airplane or a Space Shuttle. The astronaut/pilot will have to adjust his traffic pattern according to how he enters the atmosphere. The teacher will study the traffic patterns and tape them on the floor with masking tape. The children can use a model Shuttle or airplane to follow the traffic patterns, learning directionality.
    Materials: masking tape, flight patterns, model Shuttle or plane

3. The Shuttle's flight controls are like an airplane. The teacher should show the picture of an airplane to the children and ask them to find the parts that are also on the Shuttle. The children can point to or color the laminated folder game one part at a time as the parts are located. While they are doing it visualization is enhanced.
    Materials: Airplane pattern

4. To further increase visualization the matching aircraft folder game can be made by the teacher for the children to play
    Materials: matching aircraft folder game

5. A magnetic compass is a very important instrument in an airplane. It points to the North so the pilot will know which direction he is going. There is not enough gravitational pull in orbit for the Shuttle pilot to use a magnetic compass. He must use other ways to locate his position. Assemble a variety of small objects and let the children test them with the magnet. Have them sort them as magnetic or not magnetic. Classification skills are enhanced.
    Materials: Magnet, various small objects

6. The airplane maze and dot-to-dot airplane folder games can also be played by the children to enhance numeration to ten.
    Materials: folder game patterns, markers, glue

PARENT/CHILD EXPERIENCE:

1. Try to visit a local airport with your child. If a pilot is available he may show you the parts of an airplane. Your child can compare the different airplanes at the airport.

2. After the airplane lands it may go to an airport garage called a hangar. The child can make airplane hangars out of boxes and count small model airplanes as they enter into their hangars. If boxes and model airplanes are not available the folder game pattern can be used to create this game. Numeration to ten will be enhanced.
   Materials: boxes, model airplanes or folder game patterns, scissors, markers

OBJECTIVE: Visit an airport; place airplanes in hangars

BACKGROUND INFORMATION:

After the Wright Brothers flight in 1904 until World War I in 1914 only about 1,000 aircraft existed in the whole world. During the next four years thousands of aircraft were produced to be used in combat. Mail and passenger service began in 1925. During World War II more aircraft were produced before jets emerged. Today in the United States about one million people fly small airplanes yearly. Many pilots and other aerospace jobs are necessary. The pilots that fly the Space Shuttle are airplane pilots with 1,000 hours flying jets before they apply for astronaut training. They are military pilots paid by the military according to their service rank whether they fly a small airplane or the shuttle. Some of these pilots are Colonel Frederick Hauck, Colonel Frederick Gregory, and Colonel Charles Bolden. Pictures and information about pilot astronauts can be obtained free by writing NASA. (See Resources). Information about the military pilots organization, Daedalians can be obtained from Kelly Air Force Base, Bldg. 1660, San Antonio, Texas 78241. Awards, trophies and youth scholarships are presented to those who work toward the Daedalian goals, i.e. insuring America's preeminence in Air and Space. Upon request, local Daedalians may visit your class to tell you about becoming a pilot.

HOW DOES THE PILOT KNOW WHERE TO GO?

A pilot has a map that shows airports and mountains and rivers and railroad tracks and other landmarks that you can see from the air. He has a radio to talk to people who tell him where to fly and how high to fly. He also has a compass to tell him what direction to go. A compass points to north, south, east and west. You can make a compass with a dish of water, a cork, a needle and a magnet.
Rub a needle in one direction against the magnet. Put the needle through a cork. Lay the cork in a pan of water. The needle will point to the magnetic north.
An astronaut pilot can not use a compass because there is not enough gravitational pull for it to work. He uses the system of star tracking — he locates a star and follows it.

FLIGHT PLAN

Departure Point

Departure Time

Destination
(airport, city)

Estimated time enroute

Pilots name

Address

Telephone

Number aboard

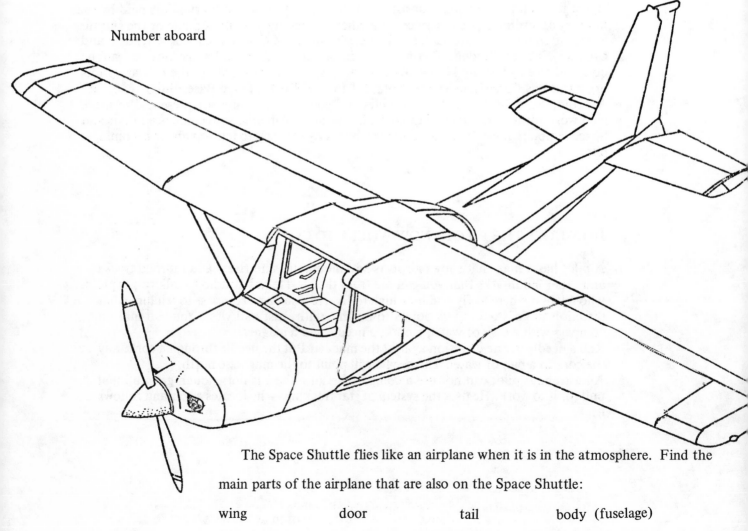

The Space Shuttle flies like an airplane when it is in the atmosphere. Find the

main parts of the airplane that are also on the Space Shuttle:

wing   door   tail   body (fuselage)

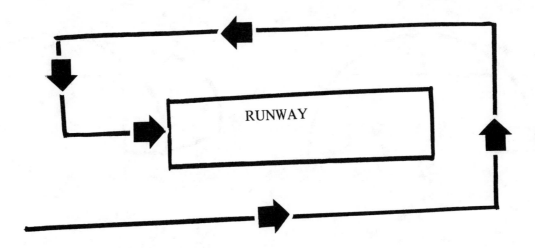

STANDARD TRAFFIC PATTERN

Pretend you are a pilot landing your plane. Listen to the planes as they take-off and land. Sounds (loud-soft, vibration, caused by aircraft noise and the direction from which the plane is coming or going) will be enhanced.

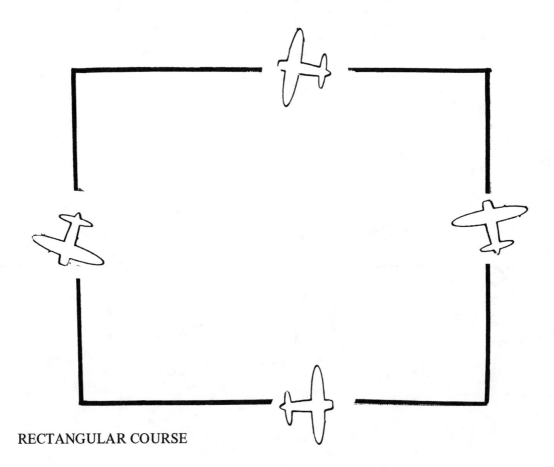

RECTANGULAR COURSE

A pilot-astronaut follows a flight pattern as he lands on the earth. He must vary the flight pattern according to different information he receives from Mission Control and his flight instruments.

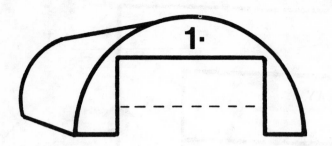

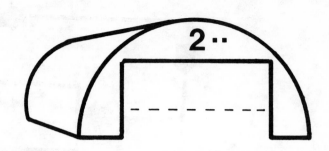

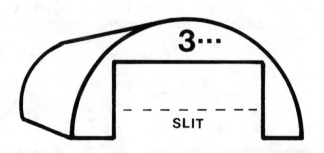

SLIT

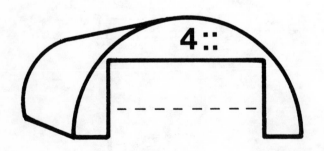

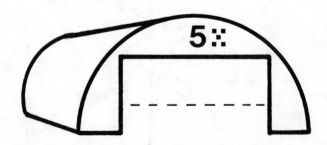

48

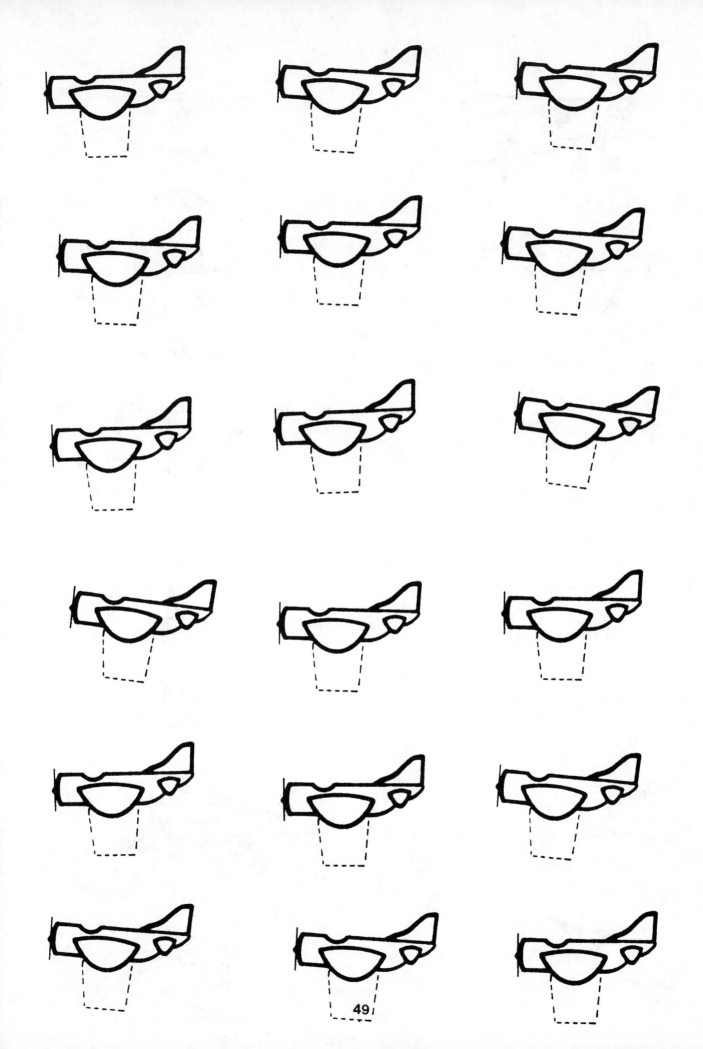

49

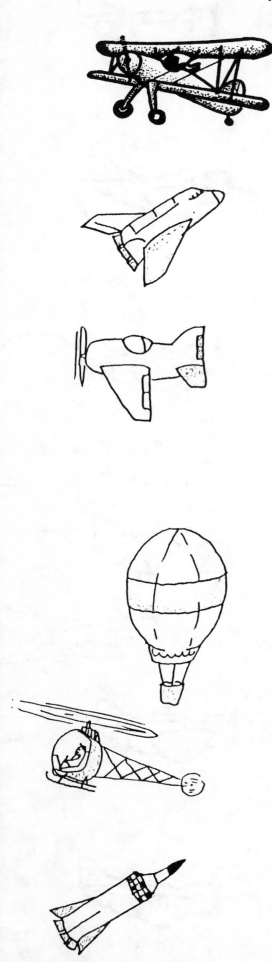

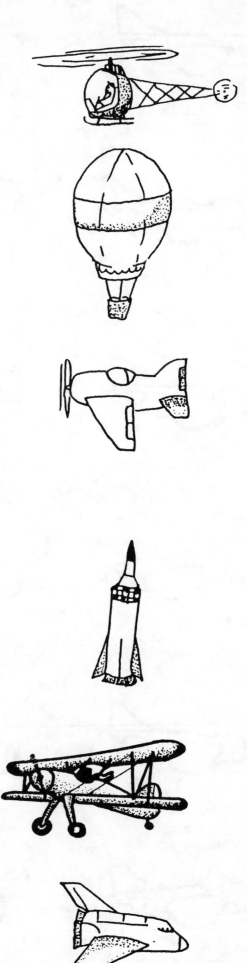

AIRPLANE

52

Airport

AEROSPACE   JOBS

NAME: Aerospace Jobs

SUBJECT/SENSE: Science (Hearing, sight)

SKILL: Become aware of different jobs

PROCEDURE:

1. The teacher will discuss some of the peopole that work at an airport using the free publication.

   A Trip to the Airport, FAA
   800 Independence Ave., S.W.
   Washington, D.C. 20591
   Read the story. (It is in English and Spanish)
   Materials: Publication

2. The teacher can invite an airport worker to class to discuss his job.
   Materials: Guest speaker

3. The teacher can discuss some of the jobs that the children are familiar with that can be performed in aviation-space environments. For example:

   Policeman — security is very important at the airports and around the Space Shuttle to prevent accidents.
   Fireman — airplanes and rockets may catch on fire because of the intense heat and huge amount of fuel, so firemen are very important
   Artists need to design and draw new airplanes and spacecraft and space stations based on principles of flight.
   Builders are needed to make the planes and spacecraft.
   Mechanics are needed to fix the planes and spacecraft.
   There are many, many more jobs.
   Materials: Prop boxes with various items to help the children role-play. These jobs can be placed in the drama center

4. Make pilot head sets or use available earphones with a mouthpiece attached. Encourage the children to talk into the earphones dramatizing being a pilot.
   Materials: Headset, mouthpiece

PARENT/CHILD EXPERIENCE:

Encourage your child to tell about an imaginary airplane trip. Encourage him to tell about where he is going, what he would find, and why he wants to go there.

OBJECTIVE: Talk about an airplane trip

BACKGROUND INFORMATION:

If possible take a field trip to an airport to observe how many different types of jobs you saw, or read the FAA Career Awareness Series to provide more background on aerospace careers. (See Resources) Use the aerospace symbols for name tags.

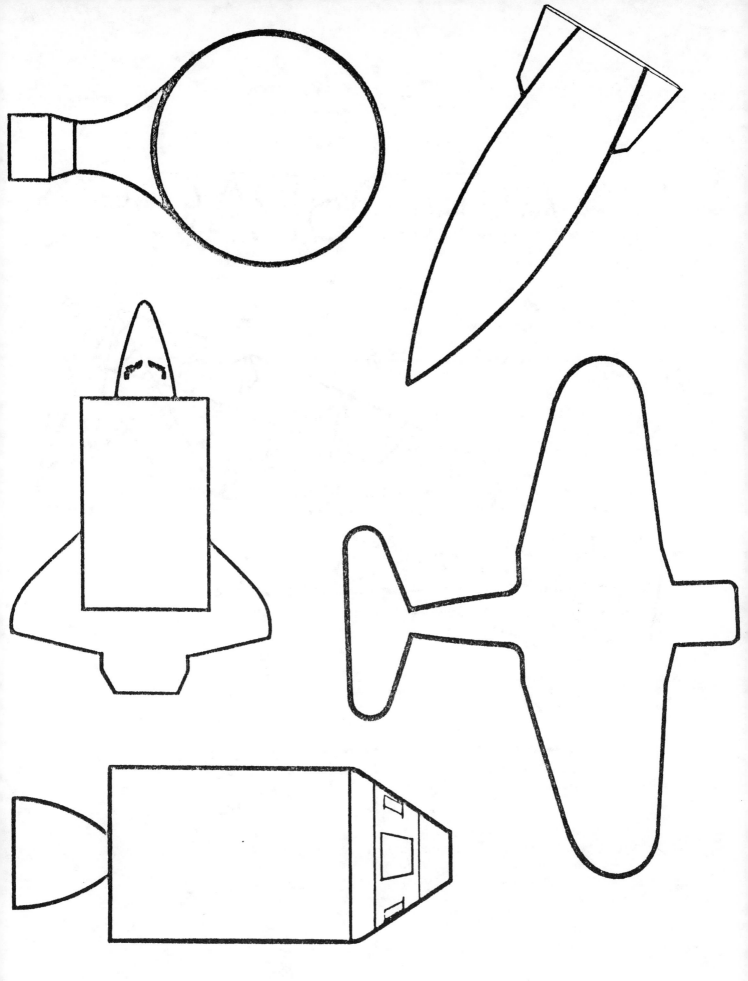

AEROSPACE

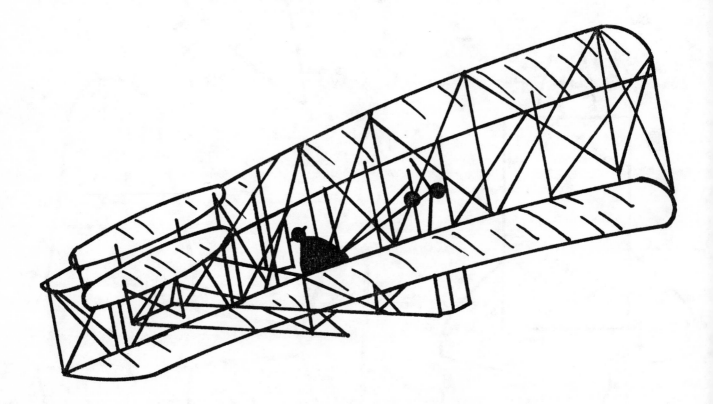

PIONEERS

NAME: Aerospace Pioneers

SUBJECT/SENSE: Science (Hearing, sight)

SKILL: Become familiar with role models

PROCEDURE:
1. The teacher will read the story about the Wright brothers to the children and ask questions i.e., Who made the first flight? Where did the Wright brothers make their first flight? Listening and comprehension will be enhanced.
   Materials: folder game stories

2. Show pictures of other aerospace pioneers. Talk about them. Obtain free astronaut pictures from NASA.
   Materials: Astronaut pictures (See Resources)

3. Many people in aerospace can become role models to children as they grow older. Talk about different pioneers in aerospace.

PARENT/CHILD EXPERIENCE:
Talk to your child about aviation when you were his age. How did you travel compared to how people travel today?

OBJECTIVE: Listen and answer questions

BACKGROUND INFORMATION:
Christa McAuliffe, the first teacher selected to go into space was going to present the lesson "Where We've Been, Where We're Going, and Why." We must know something about pioneers in aerospace so we can provide role models for children.

1792 — Montgolfier's built the first balloon
December 7, 1903 — Wright Brothers built first powered sustained and controlled flight of heavier than air vehicle
1914-1934 — Charles Lindbergh, Amelia Earhart, Jimmy Doolittle set speed and altitude records
1947 — Chuck Yeager broke the sound barrier
1962 — John Glenn, First American to orbit the earth
July 20, 1969 — Apollo 11's lunar module Eagle, with Neil Armstrong and Buzz Aldrin, landed on the moon.
1981 — Space Shuttle Columbia successfully launched with John Young and Robert Crippen
(See astronauts for update of 51-L)

Orville and Wilbur Wright were two brothers that built bicycles. They experimented in their bicycle shop and built the first airplane with a power motor. Then it came time to try to fly the airplane. As the sun rose on December 17, 1903 in Kitty Hawk, North Carolina, a bitter wind blew. The brothers took off! The airplane worked! They had done something man had always dreamed of doing but never had — fly!

The people in the Wright Brother's neighborhood did not think it was a very important thing that happened, but we know it was. If it weren't for that first flight a million people a day would not fly on airplanes today.

LEARN MORE ABOUT "99's", WOMEN AEROSPACE PIONEERS

KATHERINE STINSON was the fourth woman in America to become a licensed pilot, receiving her permit from the Aero Club in 1912, at the age of 19 years. Her instructor was Max Lilgenstrand, the legendary Max Lillie, who instructed so many early pilots. She was authorized to fly the mail at a state fair in Helena, Montana in late 1913, thus becoming the first American woman to do so. She became the first woman to loop-the-loop on July 18, 1915, over Grant Park, Chicago. She was the first woman to do night sky writing on December 17, 1915, before a large crowd of newspapermen in Los Angeles, when she traced the abbreviation for California with a magnesium flare attached to her plane. In 1916, she went on an extended tour of Canada, Japan and China, performing before record crowds. In 1917 she set a distance and duration record of 610 miles between San Diego and San Francisco, which had taken 9 hours and 10 minutes, an unheard-of feat. In May of 1918, she attempted a mail flight from Chicago to New York, making an emergency landing on a hillside in Binghamton, N.Y. However, she had been in the air 10 hours and 10 minutes, and the Aero Club awarded her the new duration record. She was the first woman to own a flying school, where she, her sister Marjorie and brother, Eddie, taught Canadians for the Royal Air Force of WWI. Katherine was forced to retire from active flying in 1920, because of health reasons. She settled and remained in Santa Fe, New Mexico until her death in 1977.

AMELIA EARHART first came into prominence in June, 1928, by being the first woman to cross the Atlantic in an airplane. Although she was a licensed pilot, she went as passenger with Wilbur Stultz, pilot and Lou Gordon, mechanic. In August, 1929, she finished third in the first Women's Air Derby. In November, 1929, she set a world's speed record for women of 184.17 mph. This was followed by three more records for speed set in 1930, and one for altitude in 1931. In May, 1932, she flew alone from Newfoundland to Ireland, a distance of 2016 miles, which took 15 hours and 18 minutes — in a single engine Vega. In July, 1932, she made a cross-country from Los Angeles to Neward, both setting records. She was to best this record, over the same route in July, 1933. In January, 1935, she made the first solo flight from Honolulu to Oakland, and in April, made a solo flight from Burbank to Mexico City. This was followed in May by a non-stop flight from Mexico City to Newark. Her last flight began in May, 1937 and ended July 2, 1937, somewhere between Lae, New Guinea and Howland Island. Her disappearance is one of the great mysteries of aviation as there has never been any trace of Amelia, the Lockheed Electra, nor Fred Noonan, navigator.

JACQUELINE COCHRAN still remains America's best known pilot, with the exception of Amelia Earhart. Jackie was born into abject poverty and spent her early years with a foster family on the sawdust trail of Florida's western panhandle, living a nomadic existence with the itinerant woodcutters and millhands. With almost no formal education, Jackie would rise from the squalor of her early life to become one of the most well known and accomplished women in the world. At the age of eight, she struck out on her own. She chose her name from the telephone book and bought her first pair of shoes with money she earned pushing heavy carts and running errands for 12 hours a day in a Georgia cotton mill. She became a beautician and developed her own line of cosmetics, which she promoted from coast to coast, flying her own plane, a Travelair. Jackie won the Bendix Trophy in 1938. In 1941, she was authorized to train and direct a Woman's Auxiliary Service Pilots (WASP) program in this country and by the war's end 1000 women pilots were fully trained and on duty. Jackie was the first woman to break the sound barrier and over a period of years she set literally hundreds of speed and altitude records, and was the recipient of the Harmon Trophy fifteen times as out-standing woman flier of the world, and became the first living woman to be enshrined in the Aviation Hall of Fame (1971). Jackie was forced to give up flying in the early 1970's because of a heart condition, and made her last flight to new horizons on August 9, 1980. Jackie was a living legend, a rags-to-riches heroine of the sort created only in America and one of the bravest women of all time.

FAY GILLIS WELLS heads up a long list of "firsts." She was a charter member of the Ninety-Nines, present at the first meeting at Valley Stream, clad in a pair of cover-alls. Fay took her first flying lesson on August 6, 1929, and soloed after logging twelve hours of instruction. On September 1, 1929, she was forced to parachute from her disabled ship after the tail collapsed, thereby becoming the second woman member of the Caterpillar Club. As a resident of the Soviet Union, 1930-34, she became the first American woman to fly in Russia and also, the first woman to own and fly a glider there. Fay became the first "air saleswoman," when she accepted a job with Curtiss Flying Service in September, 1929. While living in Russia, she greeted many of the famous flyers on stop-over there and duly recorded them as a correspondent for the Associated Press. In the spring of 1935, Fay was to accompany Wiley Post across the Bering Sea, but eloped with the famous foreign correspondent and flyer, Linton Wells, and elected to to on assignment with him to cover the Italo-Ethiopian war, thereby being replaced by Will Rogers. He would perish along with Wiley Post when their plane crashed on take-off from Barrow, Alaska. Fay retired from flying to raise a family, moving to Washington, D.C. in 1963, to become White House correspondent, covering the worldwide travels of four presidents. Fay is still active in aviation endeavors, such as the Forest of Friendship in Atchison, Kansas, dedicated to those who have contributed to all facets of aviation.

NANCY-BIRD WALTON, who obtained her pilot's license 1933, became the youngest female commercial pilot in the British Empire, indulging a passion for flying which manifested itself when she was 13. She was trained as a pilot with ''Smithy'' himself, Sir Charles Kingsford-Smith. In an effort to make a financial career out of aviation, she began with a barn-storming stint around New South Wales in the 1930's. In 1935 she organized the First Ladies' Flying Tour in Australia and flew approximately 22,000 miles and carried 900 passengers. She was owner-pilot to the Far West Children's Health Scheme, flying aerial ambulance and private charter work. In 1940, Nancy-Bird conducted a survey of women pilots with over 500 hours experience and able to go to England to serve in the Air Transport Auxiliary but they were refused permission to leave Australia. Nancy-Bird Walton is known as the first lady of aviation in Australia, still actively flying and promoting aviation. She is the patron of the Australian Women Pilots Association, has received Australian recognition with the award of an OBE.

SALLY RIDE was one of the six highly educated women who arrived on America's newest frontier, space travel, in 1978. She was the first American woman in space, flying as flight engineer on the Challenger in 1984. Her arrival at the Johnson Space Center in Houston, along with the other five women astronaut candidates liberated the most exclusive men's club in America — no area of national life had been so totally male as the space program. Sally wasn't chosen by NASA because she was a woman, she was chosen because she was well qualified. ''Sally is a very smart lady,'' says Bob Crippen, a fellow astronaut, ''And she's smart in a special way. You get people who sit in a lab and think like Einstein, but they can't DO anything. Sally, however, can get everything she knows together and bring it to bear where you need it.''

The history of women pilots in the U.S. scheduled airline industry began in January of 1973, with the hiring of **Emily H.Warner** by Frontier Airlines. Another industry ''first'' came in June of 1976 when Ms. Warner was elevated to the rank of captain, commanding a deHavilland Twin Otter prop-jet. In 1973 she received the Amelia Earhart Award as the year's outstanding woman in U.S. aviation. Later that year, she was presented Colorado's first Wright Brothers Memorial Trophy. In 1974, she was the first woman elected to membership in the Air Line Pilots Association. On May 4, 1984, Emily Howell was made Captain for Frontier Airlines, flying a Boeing 737.

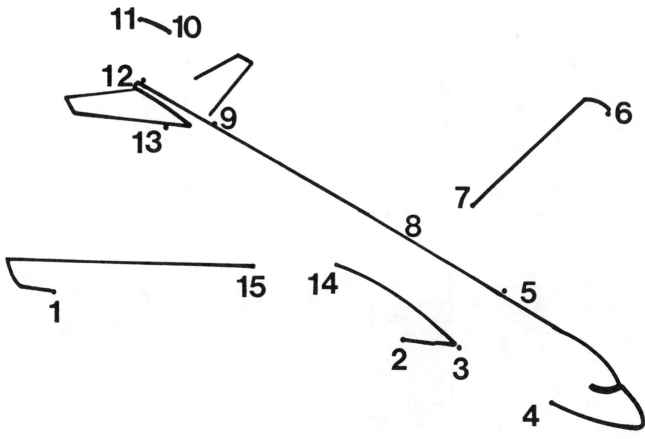

★ ★ ★ ★ ★ ★ ★ ★ ★ <ROCKETS<<

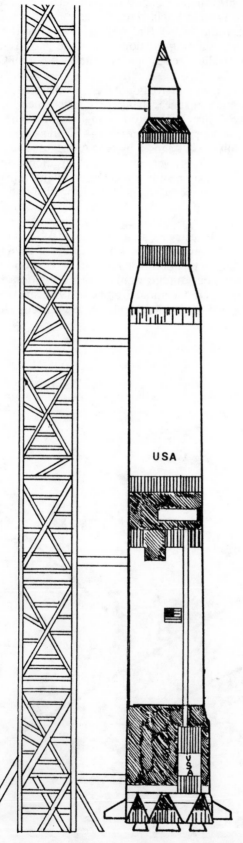

USA

NAME: Rockets

SUBJECT/SENSE: Math (Sight) Science (Sound)

SKILL: Become aware of thrust; reinforce sequential ordering

PROCEDURE:

1. The teacher can cut out the rocket sequence pictures, color and laminate. Place numerals on the back to make the game self-correcting. The children can arrange in sequential order. Younger children will use less pictures and work up to the six frames. (The pictures can be attached at the top and flipped at the bottom to create a flip book.)
    Materials: folder game sequence pictures, scissors, colors

2. The teacher can cut out the rockets and rectangles and laminate for durability. Seriation (putting objects in order) can be introduced. The children can place the rockets in a series from short to tall and tall to short. This concept will be extended with the ten rectangles.
    Materials: folder games seriation patterns, scissors

3. The teacher will cut a 6-foot length of string. Thread it through a straw. Tie each end of the string to the back of a chair or some other object which will keep the string stretched tight. Next, blow up a balloon and, while keeping it tightly closed to prevent the air from escaping, tape the balloon to the straw (see diagram). Release the balloon. Measure how far it went. Repeat this three times, using different amounts of air each time. Ask the children what they observed about the distances traveled.
    Materials: string, straw, balloon

PARENT/CHILD EXPERIENCE:

The parent will blow up a balloon to demonstrate that a rocket needs thrust to blast-off. As your child lets a balloon go the air is released from it causing it to move forward.
Materials: balloon

OBJECTIVE: Listen to the rocket balloon

BACKGROUND INFORMATION:

A rocket is based on Newton's Law that for every action there is an opposite reaction.

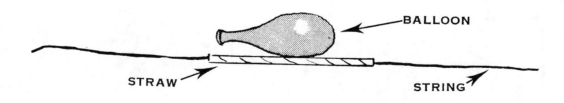

BALLOON

STRAW STRING

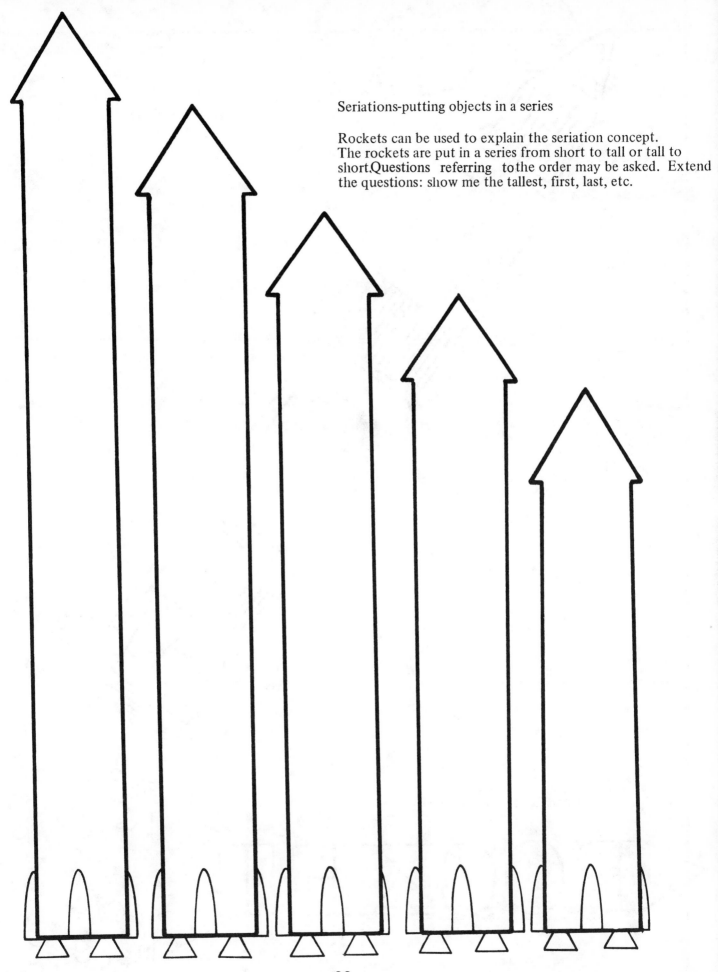

Seriations-putting objects in a series

Rockets can be used to explain the seriation concept.
The rockets are put in a series from short to tall or tall to
short. Questions referring to the order may be asked. Extend
the questions: show me the tallest, first, last, etc.

63

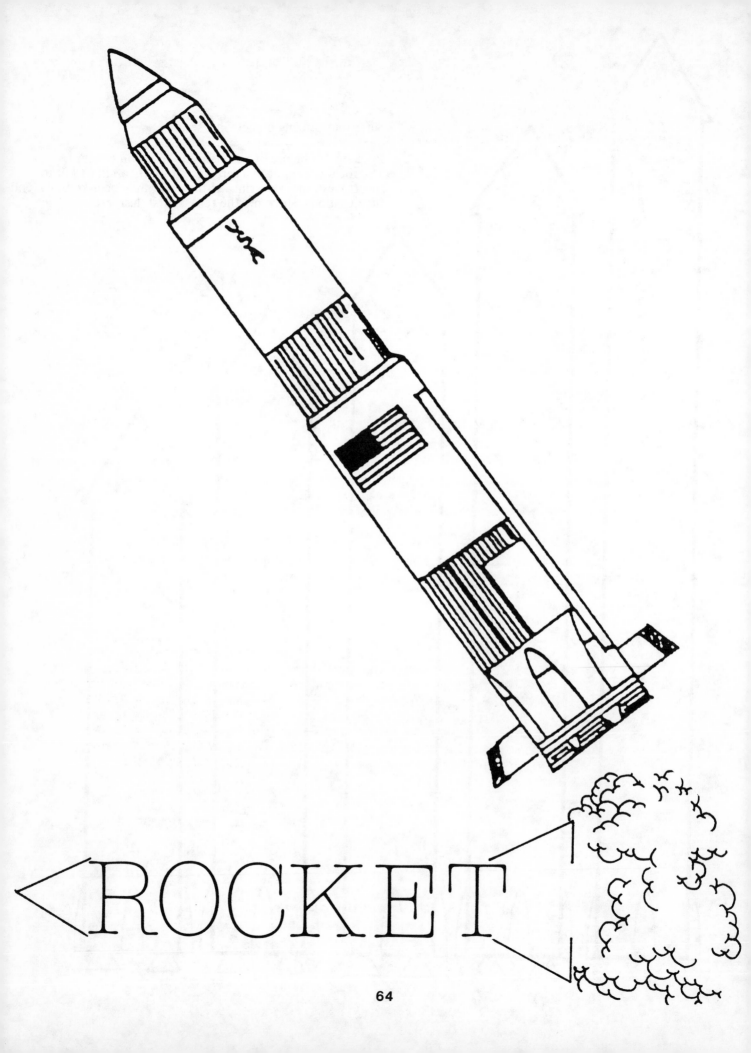

ROCKET

NAME: Rocket

SUBJECT/SENSE: Science (Hearing), Math (Hearing)

SKILL: Enhance shape recognition, develop fine motor development, and enhance auditory discrimination

PROCEDURE:
1. The teacher will read the rocket poems. The children will follow along acting out a rocket blast-off.
    Materials: poems

2. The teacher will laminate the rocket tangram (shapes reassembled into different figures). Laminate a second set of rocket shapes or trace onto cardboard. The children can match the squares and triangles to the rocket shape. The shapes can be color-coded for younger children to match shape and color.
    Materials: folder, pattern, glue, scissors, markers

3. Show different pictures of rockets to the children. Encourage them to build rockets and launch pads with their blocks and dramatize a rocket launching.
    Materials: blocks, rocket pictures available from NASA (See resources)

PARENT/CHILD EXPERIENCE:
Help your child make a rocket out of a cardboard cylinder (see next page for directions).
Materials: cardboard cylinder, paper, scissors, glue or tape

OBJECTIVE: Create a rocket

BACKGROUND INFORMATION:
After the launching of Sputnik, NASA was established to manage our new space program. Liquid fueled rockets launched satellites into orbit around earth. The Mercury Program (1961-1963) with six missions and the Gemini Program (1965-1966) with ten missions preceded the Apollo Program (1968-1972) with 11 missions. During the Apollo Program a giant rocket called Saturn 5 took men to the moon. The Shuttle program uses solid-fueled rockets to assist the liquid-fueled rockets to take the Shuttle into orbit. The solid-fueled rockets fall into the ocean and are recovered. The liquid-fueled rockets bring the Shuttle back to earth. (There were 24 missions prior to the Shuttle disaster.)

ROCKETS

Procedure   The simplest rocket that you can make is with a toy balloon. When the balloon is inflated the pressures acting against the wall of the balloon are in balance. When the outlet is opened, gas discharges through the opening and the balloon moves in the opposite direction. This is the same principle of the actual rocket engine. Why does the balloon (rocket) move forward? Because of Newton's law: For every action there is an equal and opposite reaction.

Materials   Empty cardboard cylinders
Rolled up paper, paper cup
Tape
Paint
Any other materials the children can think of. Let them be creative!

Procedure   Have the children study pictures of rockets and then design their own. They may like to decide where they will blast off to and what they will discover.

Procedure   Have the children learn the poems below.

I'M A LITTLE ROCKET (SHUTTLE)

I'm a little rocket (child squats)
Pointing to the moon (points arms upward)
4...3...2...1... (repeat slowly)
Blast off! Zoom! (spring into the air)

DID YOU EVER SEE A ROCKET
(tune-Did you ever see a Lassie)

Did you ever see a rocket,   a rocket, a rocket?
Did you ever see a rocket go this way and that?
Go this way and that way, go this way and that way,
Did you ever see a rocket go this way and that?
(repeat using aerospace terms)

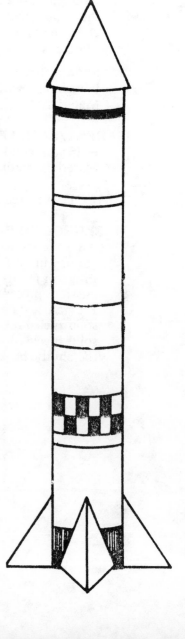

Laminate the tangram rocket shape onto the
file folder. Cut out a second set of shapes.
Allow the child to match the shapes. For younger
children shapes can be colored to match.

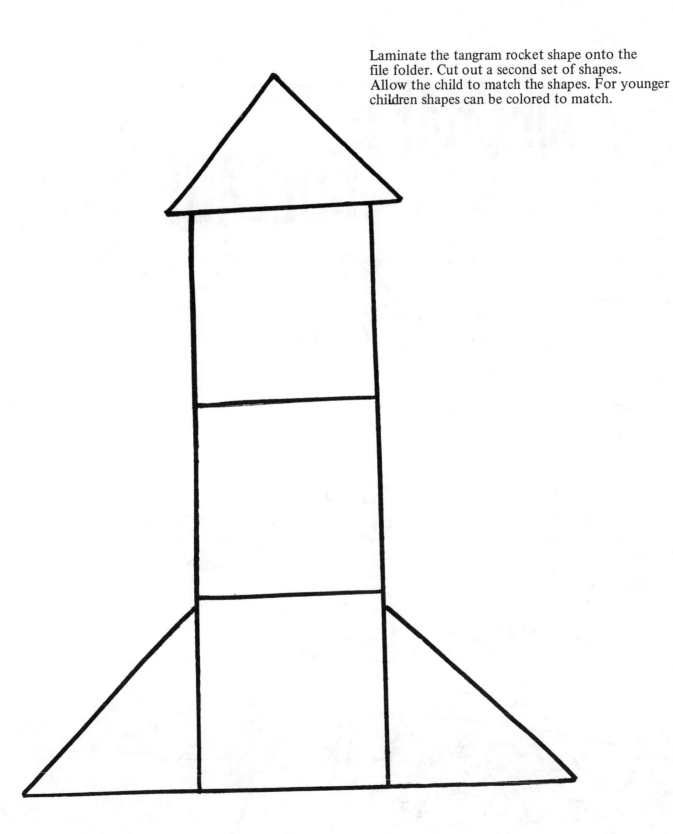

AIRCRAFT COLLAGE

68

NAME: Aircraft Collage

SUBJECT/SENSE: Math (Sight)

SKILL: Develop ability in visualization and classification

PROCEDURE:
1. The teacher can color and laminate the collage on the folder. Then ask the children to name the things that fly. This will increase the childrens' revisualization skills and reinforce their awareness of aircraft.
    Materials: folder game activity, colors, markers

2. The teacher will color and laminate the cards of Things That Fly and Things That Do Not Fly. Then she will talk about what the objects represent and let the children classify accordingly.
    Materials: folder game, patterns, scissors, markers

PARENT/CHILD EXPERIENCE:
The parent will help your child cut out magazine pictures or draw pictures of things that fly. Using a variety of media (colors, paint, drawings, pictures, etc.) encourage your child to make the collage as creative as possible. The child's concept of things that fly will be enhanced.
Materials: magazine, paper, scissors, markers, available drawing tools

OBJECTIVE: Create a collage

Note: Many inexpensive plastic astronauts and space creatures are available commercially, i.e. Star Patrol with Xwing Rocket, No.1114 Mfd. by Processed Plastic Co., Montgomery, Ill. 60538 ($1.98 for 38 figures). The children can sort and classify by color, appearance, etc., as well as use them in imaginative play and as board game pieces.

CLASSIFICATION

The child looks at the picture card. He decides which group that it goes with. He puts the card into that set. Symbols or words on the back of the card correspond to the correct envelope to make the game self-correcting.

THINGS THAT FLY   THINGS THAT DO NOT FLY

TRIP TO THE MOON

NAME: Trip to the Moon; Blast-Off

SUBJECT/SENSE: Math (Sight)

SKILL: Become aware of the moon as an object in the sky with a surface, reinforce skills

PROCEDURE:

1. The teacher can make the board game and place any skill on the game board (numbers, letters, math facts, colors, etc.) If the game board is laminated prior to placing the skills on the blank spaces, it may be used over and over, substituting the skills as needed. Markers or crayons can be used directly on the laminated surface and erased or blank stick-on labels can be used. The numbers can be cut out, placed in a cup and the children can pick a number and move the indicated number of spaces or a die may be used. The children can play the board game to see who gets to the moon first.
   Materials: folder game pattern, markers, glue

2. The teacher can teach the song "Trip To The Moon" to the tune of "She'll Be Coming 'Round the Mountain."

   They'll be coming 'round the planet when they come, zoom, zoom
   They'll be wearing their new space suits when they come, zip, zip
   They'll be riding the Space Shuttle when they come, shush, shush
   They'll be stopping at the space station when they come, stop, stop

3. The teacher will collect different types of rocks and dirt with the children. Then the children can compare the sizes, shapes and colors of the rocks. Fill one gallon plastic jug with one kind of soil (about one quart) and water (about three quarts), shake the jug and let it stand. The children can observe the soil layers as they form (rocky and sandy soil form different layers). Compare the texture, consistency and color of the dirt.
   Materials: rocks, dirt, gallon plastic jug, water

PARENT/CHILD EXPERIENCE:

Observe the moon phases at night with your child. Talk about how the moon changes shapes. Encourage your child to draw these shapes.

OBJECTIVE: Observe the moon

BACKGROUND INFORMATION:

There were 12 astronauts that landed on the moon and collected dirt and rock samples. There were six landings of Apollo (11, 12, 14, 15, 16, 17) on the moon. Many scientific studies were made. One experiment was dropping a hammer and a feather. They landed at the same time because there was no air resistance. (Teachers can obtain a lunar soil certification from NASA workshops and check out the moon rocks to show their students.)

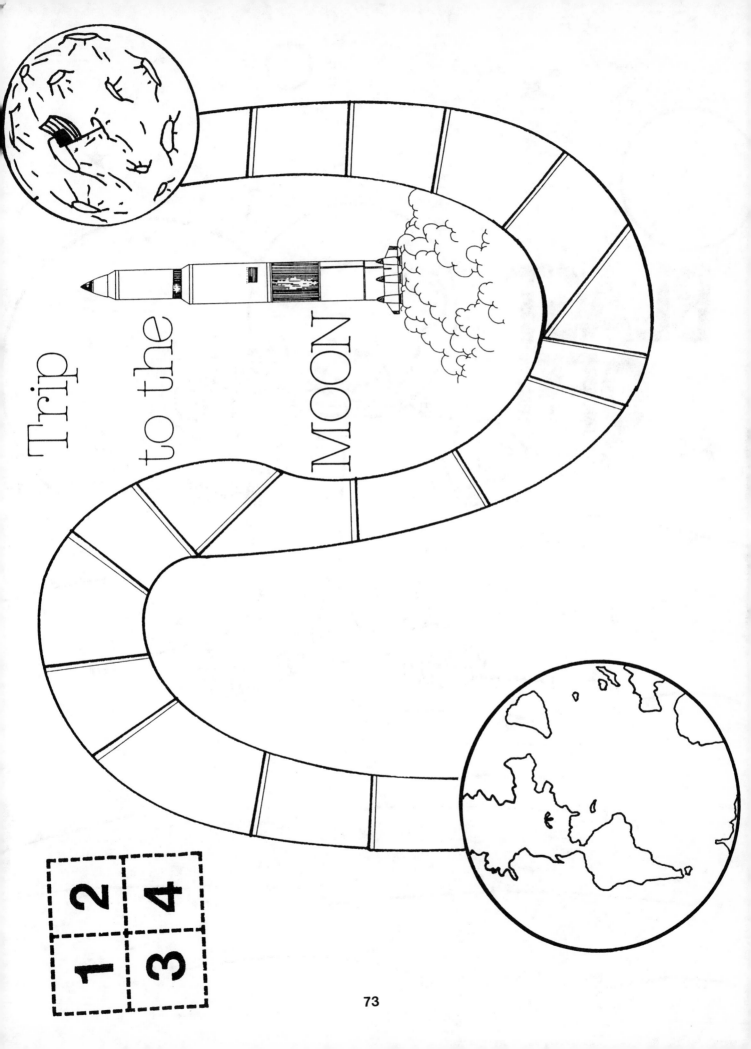

Trip to the MOON

1 2
3 4

73

BLAST
OFF!

74

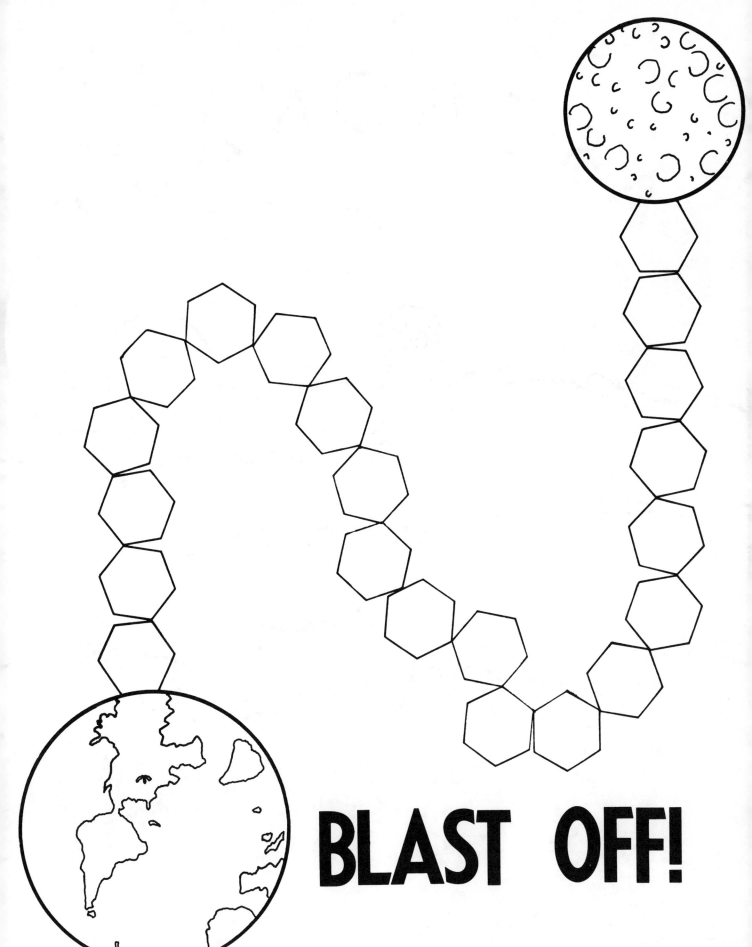

BLAST OFF!

ASTRONAUT

NAME: Astronaut

SUBJECT/SENSE: Math (Sight, hearing)

SKILL: Form sets to 10

PROCEDURE:

1. The teacher can glue the astronauts to file folders and laminate them for durability. The "Astrosticks" can be glued to popsicle sticks or a paper ring can be glued to the back to create a finger puppet. The teacher will read the poems to the children and then have them repeat. The children can use the astrosticks or astropuppets to act out the poems.
    Materials: astronaut patterns, file folder, markers, scissors

2. The teacher will make the set cards. She will color the sets, laminate and cut out. The children will match the number word with the numerals reinforcing the concept of numeration to 10.
    Materials: folder patterns, markers, scissors

3. Encourage the children to dramatize the landing on the moon. Buzz Aldrin and Neil Armstrong were the first men on the moon. Talk about this event.
    Materials: pictures of this event can be obtained from NASA (See Resources)

PARENT/CHILD EXPERIENCE:

With your child make a moon landscape. Place small objects on a piece of cardboard or other hard surface. Cover with tin foil. This will create a surface of craters and mountains. Purchase 10 small toy astronauts, space ships, etc. and encourage your child to act out the moon landings.
Materials: tin foil, cardboard, small objects, toy astronauts

OBJECTIVE: Create a surface representing craters and mountains

BACKGROUND INFORMATION:

The original 7 astronauts in the Mercury Program were: Sheppard, Grissom, Glenn, Carpenter, Schirra, and Cooper. The Gemini Program flew 20 astronauts and the Apollo Program flew 33 astronauts. There have been 132 crew members to fly aboard the Space Shuttle, although some astronauts flew more than one mission. The last astronauts selected to fly but lost their lives were: Dick Scobee, Mike Smith, Judy Resnick, Ellison Onizuka, Ron McNair, Greg Jarvis and Christa McAullife — the first teacher selected to fly in space.

TEN LITTLE ASTRONAUTS (Tune: Ten Little Indians)
One little, two little, three little astronauts,
Four little, five little, six little astronauts,
Seven little, eight little, nine little astronauts,
Ten little astronauts,

TWO LITTLE ASTRONAUTS
Two little astronauts are going to the moon.
Two little astronauts hope they'll get there soon.
The first one said, "Oh, this is such fun."
The second one said, "We will see the sun."
Then — 10, 9, 8, 7, 6, 5, 4, 3, 2, 1, ZOOM!

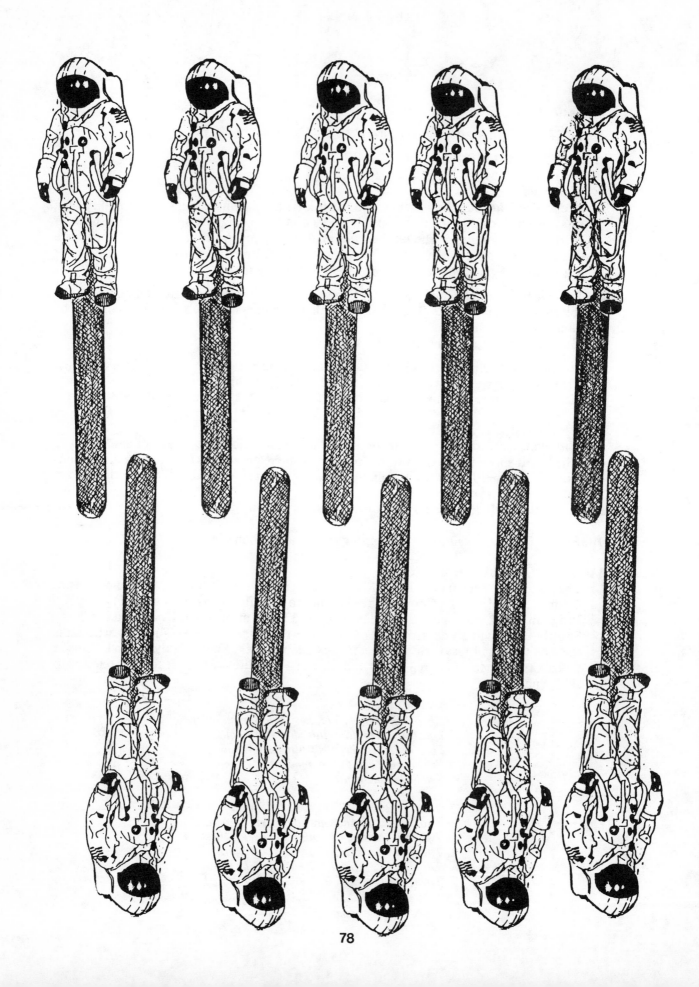

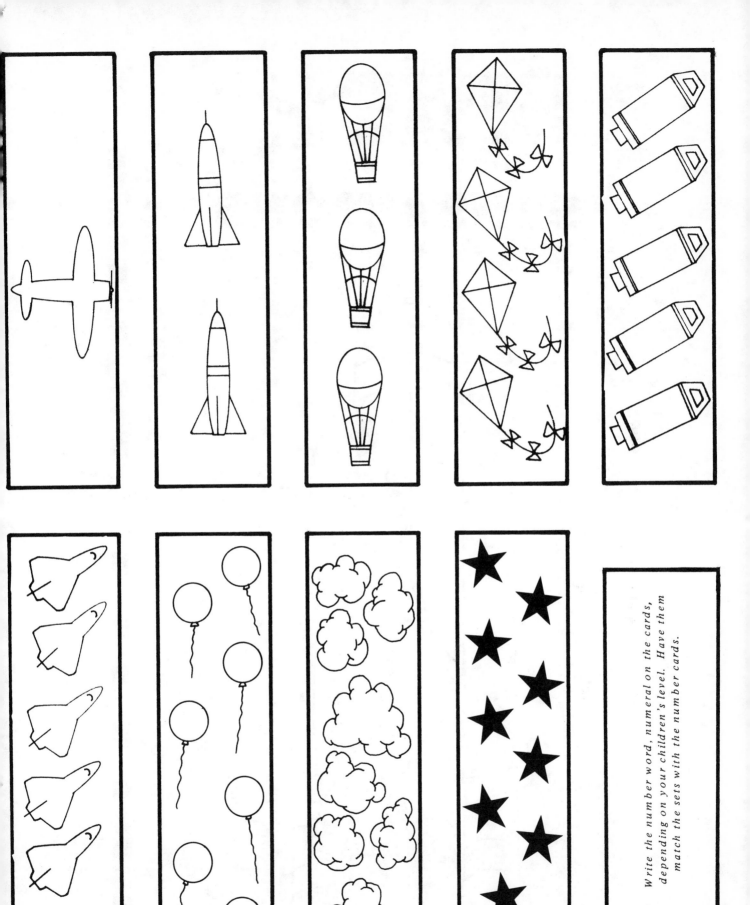

Write the number word, numeral on the cards, depending on your children's level. Have them match the sets with the number cards.

SPACE
SENSATIONS

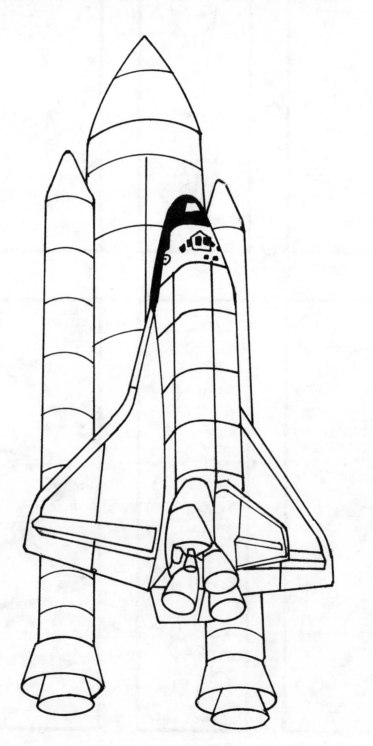

NAME: Space Sensations

SUBJECT/SENSE: Science (Taste, Smell, Touch, Hearing, Sight, Sound)

SKILL: Become aware of the senses (for safety and enjoyment)

PROCEDURE:
1. The teacher can discuss how an astronaut feels during the blast-off. The astronauts are on their backs and the Shuttle vibrates and shakes. The children can sit in a chair and the teacher can gently shake the chair to give them the sensation vibration. As the Shuttle goes into space a pressure is applied toward the body. The children can go outside and simulate this experience with a swing. As they swing forward they can feel the pressure. At the top of the swing there is a moment of weightlessness, resembling what you would feel like in space.
   Materials: chair, swing

2. Astronauts have to go through the same daily routine that they do on earth. They must rest by strapping themselves in a sleeping bag so they won't float around the Shuttle. They can sleep in any direction since there is no gravity. On earth we must lay down to rest the heart. At rest time lay a strip of paper across the children. As they lay on their mat or towel have them pretend they are sleeping in space. An eye mast may also be used to block out light since astronauts see daylight every 45 minutes.
   Materials: scrap material, elastic or string attached to the ends.

3. Tape record different sounds as you and the children go for a walk. Have the children identify the sounds as you play the tape. Astronauts carry tape recordings of their favorite music to play in space. One of the astronauts favorite tapes on the Shuttle was a tape of earth sounds; birds singing; waves at the beach; leaves rustling, etc.
   Materials: tape recorder, tapes

PARENT/CHILD EXPERIENCE:
The astronauts see a sunrise and sunset every 45 minutes as they orbit the earth. An astronaut's day is the same as a day on earth — there are still 24 hours in a day, however, we see the sunrise and sunset at different times. With your child, watch the sunrise and sunset. The child will learn the day (sun) and night (dark) concept.

OBJECTIVE: Observe a sunset and/or sunrise

BACKGROUND INFORMATION:
Astronauts brush their teeth and bathe in space differently than on earth. They can swallow their toothpaste or spit it in a towel, but they can't spit it into the sink. They wash their hands and faces with sudsless or regular soap and wet a washcloth from a pressurized water hose.

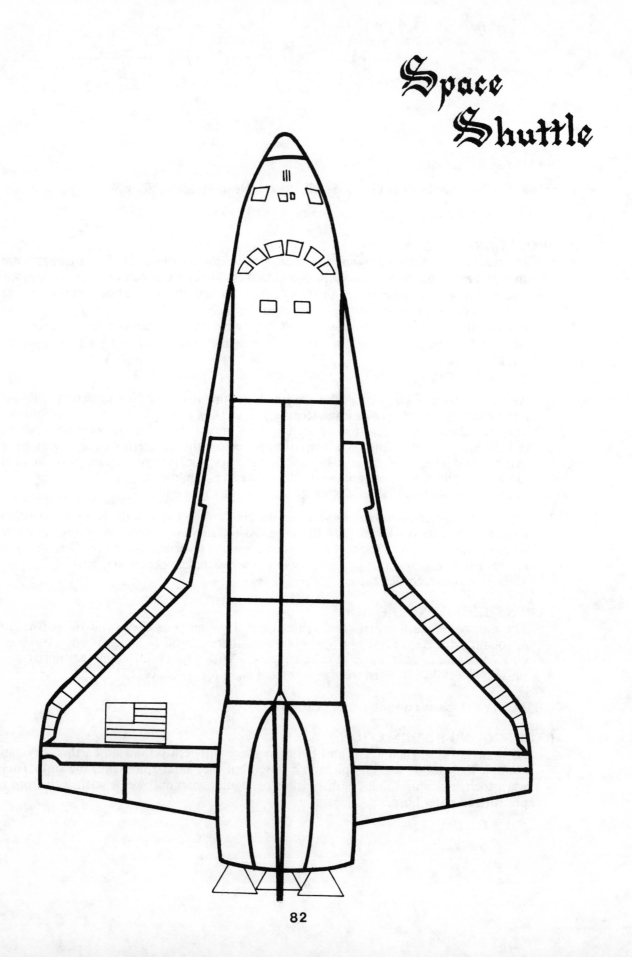

Space
Shuttle

NAME: Space Shuttle

SUBJECT/SENSE: Math (Sight0

SKILL: Become aware of limited space

PROCEDURE:

1. The teacher will help the children build a Space Shuttle in the classroom. Let the children take turns working or playing in the shuttle. They will learn about working in a confined space.
   Materials: Space Shuttle mock-up

2. The teacher can make the dot-to-dot Shuttle folder game, laminating the surface so it can be used as a wipe-off card. The children can use a crayon to complete the Space Shuttle dot-to-dot folder game learning the numerals to 11. This will encourage the children to draw a space Shuttle and create their own drawings of a spacecraft. You can also show the children different pictures of the Shuttle.
   Materials: folder game shuttle, crayons, Shuttle photos (See NASA Resources)

3. The Shuttle orbits (goes around the earth) about 180 to 250 miles above the earth. From space we can see the earth. The earth will be brown and green and the water (oceans, lakes) will be blue. White clouds would cover parts of the earth and water. You can even see the contrails (white trails) of airplanes and wakes of ships. The teacher can discuss this concept with a child by looking at a globe. Cotton can be glued to simulate the white.
   Materials: globe, cotton, glue stick

4. As the children look at a globe, they can draw pictures of the earth, as they imagine it would look like to an astronaut from space. Such songs as the following could introduce the concept of this being a small world.
   Materials: 'We Are The world' (commercial); 'It's a Small World' Walt Disney Productions, U.S.A., Western Publishing Co., 1968; 'Travelin' With Ella Jenkins' New York: Folkways Records and Service Corp., 1979. (Ella introduces the word 'hello' in several languages).

5. Children can assemble, decorate and have contests flying Thunder Bee Gliders (mini shuttles @ .30) Request order form from GBB Thunder Bee Gliders, Grivno Acres, So. Minn. St., P.O. Box 694, Crookston, MN 56716 (218) 281-4690.

PARENT/CHILD EXPERIENCE:

The parent can set up a tent at home and let your child participate in various activities — eating, paper work (coloring, etc.), playing, etc. Your child will learn about living in a confined space. Compare this to living and working in Space Shuttle.
Materials: Bed sheet

OBJECTIVE: Create a tent and become aware of limited space

BACKGROUND INFORMATION:

While the Apollo program was in progress, a Space Transportation System of which the Shuttle is the manned portion, was developed. Most of the expensive parts are returned to earth, thus allowing more flights at a lower cost. The Shuttle comes back to earth where the early spacecraft could not. Tell the children when they grow up it will be possible to travel in space to space stations and maybe to other planets.

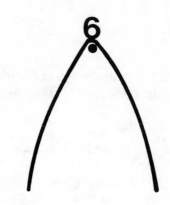

6

7•   •5

8•   •4

9•   •3

10•   •2

11•   •1

SPACE SHUTTLE MODEL

ASSEMBLY INSTRUCTIONS

Read carefully before assembly:

1. Cut out all parts using scissors.
2. Cut out V-shaped notches on Fuselage to create tabs along the outside edge. Fold tabs out.
3. Glue or tape three Nose Weights to the underside of the nose of your glider. Use the fourth weight provided if needed for extra trim after assembly.
4. Fold Fuselage along middle line.
5. Starting at the nose, glue or tape Fuselage to Deck and Wing Assembly. Match tabs on Fuselage exactly to those printed on Deck and Wing Assembly.
6. To close the nose, glue or tape the two halves together using tabs provided.
7. Fold Vertical Stabilizer Assembly. Fold out tabs A and B. Glue or tape the Vertical Stabilizer assembly to make one solid piece except for tabs A and B.
8. Attach Vertical Stabilizer to Fuselage, matching tab A with point A and tab B with point B.

PREFLIGHT INSTRUCTIONS

For best results, launch your Shuttle glider with a gentle, level toss. Bend the Body Flap up slightly for a greater lift.

Glue pattern onto file folder.

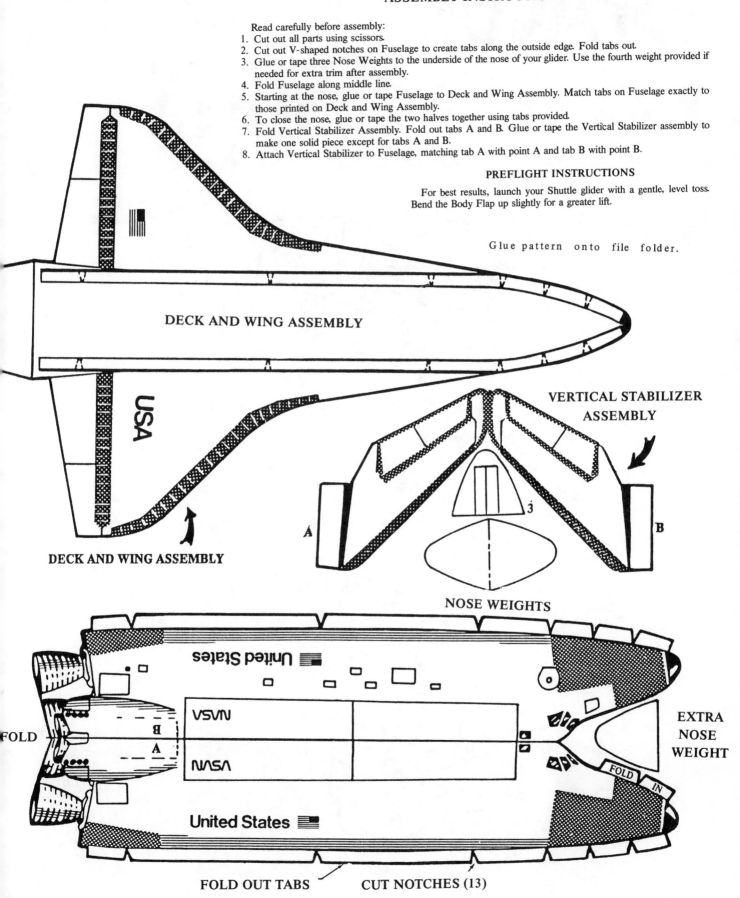

DECK AND WING ASSEMBLY

DECK AND WING ASSEMBLY

VERTICAL STABILIZER ASSEMBLY

A

3

B

NOSE WEIGHTS

United States

USA

NASA

FOLD

B

A

EXTRA NOSE WEIGHT

FOLD IN

United States

FOLD OUT TABS   CUT NOTCHES (13)

SPACE SUITS

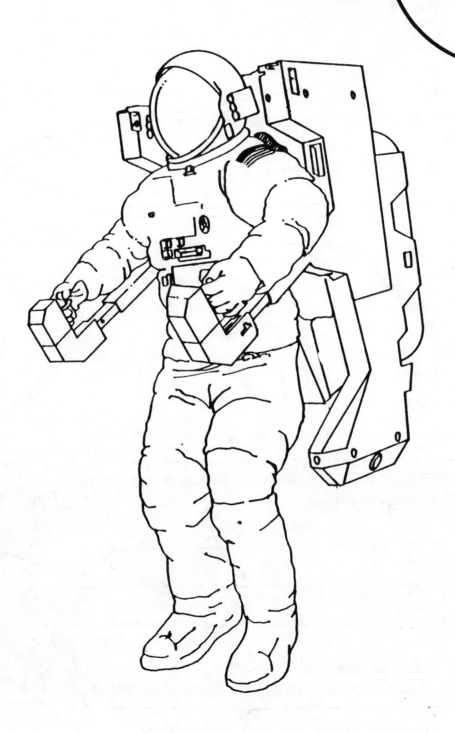

NAME: Space Suits

SUBJECT/SENSE: Science (Touch)

SKILL: Become aware that clothing is for protection and comfort

PROCEDURE:

1. The main difference on the astronauts clothes in the Shuttle is there are many pockets and attachments to hold their small items, such as scissors, tools, pencils, and utensils. If they sat something down, it would float away because there is no gravity! As the children pretend they are an astronaut, remind them to keep their things attached to their clothes or in their pockets. Strips of cloth one inch wide and two inches long can be sewn or safety pinned at the ends to old clothes.
    Materials: Cloth strips, safety pins or needle and thread (Velcro strips are used on the Shuttle)

2. Show the children the picture of the space suit. (This is the extra vehicular space suit worn as the astronauts leave the Shuttle to do experiments in space.) Ask them to see how many parts they can identify. Find the helmet, gloves (hands) and arms.
    Materials: folder game pattern

3. Although the Shuttle is reusable (returns to earth and goes into space again) many items aboard are disposable. Discuss the difference between reusable, disposable and recycleable and different disposable items that could be recycled. An example is a milk jug made into a space helmet. Cut out a place for the eyes and cover with colored cellophane. If you attach red cellophane the children will see their entire environment red — not just an item in the environment. Then you can attach blue over the red, creating purple and the children will learn primary and secondary colors: red and blue make purple, yellow and blue make green, red and yellow make orange.
    Materials: colored cellophane, plastic milk jug or round gallon container or paper sack

PARENT/CHILD EXPERIENCE:

Talk with your child about how layers of clothing insulate the body to keep you warm. The astronauts traveling on the moon needed insulation because there was no atmosphere and they needed to keep their body temperature constant so they would not die. To introduce the concept of insulation, let your child touch a cold object, i.e. ice, with his bare hand. Then let him touch the same object wearing a glove. Ask him to compare the feelings. Astronauts also needed an air supply since there was no oxygen to breathe. Oxygen tanks can be made from plastic coke bottles or oatmeal boxes, yarn, and plastic tubing (3 feet-1½ per tank — costs less than 50¢) Let your child dress up as an astronaut, encouraging him to use his imagination.

OBJECTIVE: Dress up as an astronaut for dramatic play

BACKGROUND INFORMATION:

People wear clothing for protection and comfort. Clothing has to be changed when it gets dirty. Astronauts wear the same kinds of clothes in the Space Shuttle that they do on earth. However, they may wear shorts and a shirt or a jump suit with long pants depending on how comfortable the temperature is to them. The Shuttle's inside temperature is about 73 degrees. Some other kinds of clothing items an astronaut will need include: socks, shoes or boots, underwear, sweater, gloves, hat.

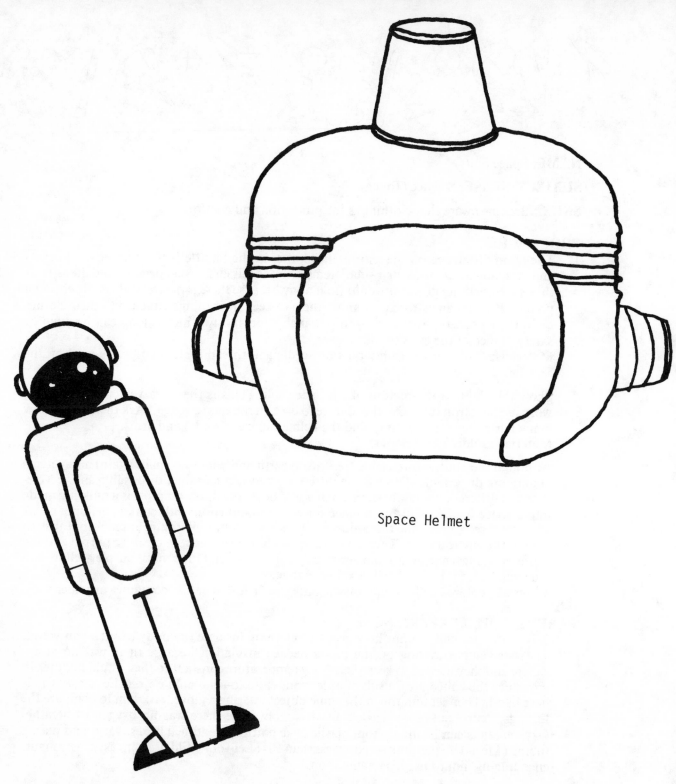

Space Helmet

Space Suit

Have children, one at a time, lay down on blue butcher paper. Trace around the body up to the neck. Cut out the body outline then draw a NASA patch, a flag, a name card, and the NASA logo on index cards and place accordingly. Have the children put their space suit in front of them with their helmets on.Take pictures for the bulletin board.

Contributed by Marjorie Shaefer, preschool teacher, Ann Arbor, MI.

SPACE
FOOD

NAME: Space Food

SUBJECT/SENSE: Science (Taste, smell, touch)

SKILL: Become aware of differently prepared foods (dehydrated, compact); become aware of heat, air and water on the texture, taste, odor and appearance of foods

PROCEDURE:

1. The teacher can talk about freeze-dried or dried foods such as Tang. Prepare Tang and compare to fresh orange juice.
Materials: Tang, water, fresh orange juice, containers

2. To demonstrate the fact that humidity is necessary to prevent dehydration the teacher can perform the following experiment. Using two containers with covers, fasten a piece of freshly peeled apple to the inside of each lid. Pour one-half cup of water into one jar. Do not put any water in the other jar. Fasten the lids on the containers and place them in the sunlight. The apple in the container without any water will dehydrate rapidly. Ask the children to say what they saw.
Materials: Two containers with covers, apple, water

3. To simulate the way astronauts eat some space foods, puree some food in a blender or make instant pudding and put in a small plastic bag. The change in consistency of some ingredients when they are mixed will be observed by the children. Let the children eat the food as astronauts do. Clip a small hole in one corner of the bag and squeeze the food into the mouth.
Materials: food, blender or instant pudding and milk, small plastic bag

4. The teacher will place small amounts of raw, canned and dried fruit in separate dishes. The children can compare the difference in a fruit that has been prepared in different ways. Taste the fruit. The children will learn the effects of heat and water on the texture, taste, odor, and appearance of foods.
Materials: raw, canned and dried fruit, dishes

5. The teacher can remove some of the fruit from commercial cereals containing freeze-dried fruit. Place the fruit in a small bowl with a small amount of water for a few minutes. Measure the water before and after rehydration. The children can observe what happens to the fruit and to the water. Ask them is the fruit still as hard as when it was first placed into the bowl of water.
Materials: cereal, water

6. Purchase space food from NASA's souvenir shops, if possible. (Contact Kennedy Space Center Visitors Center, if living in the southeast.)

PARENT/CHILD EXPERIENCE:

Have your child help prepare balanced meals at home using dehydrated baby food. Talk about the 4 basic food groups. (See parent note)
Materials: dehydrated baby food (Hines or Gerber)

OBJECTIVE: Prepare then compare taste, smell, texture of food

BACKGROUND INFORMATION:

There is no food in space so you have to take it with you. Food in space has to be dehydrated and compactly packed because at blast-off you must have the least weight possible and dehydrated food takes up less space. There is no refrigerator on the Shuttle but astronauts have fresh fruit and vegetables for as long as they stay crisp. Astronauts add water to the dehydrated food when they are ready to eat and heat the food if necessary. If the astronauts aren't careful, some food will float away. Velcro straps or magnets hold the food trays on the walls or on the astronauts' legs. Use a styrofoam meat tray with yarn attached to simulate an astronaut's food tray.

WHAT DO YOU EAT IN SPACE?

About fifty different kinds of foods are available for an astronaut to eat in space. However, he can not use a knife and fork the way we do on earth since there is no gravity. The meat might float right off his plate!

He often squeezes food out of a tube like putting toothpaste on a toothbrush. He can eat bitesized pieces of anything. A typical meal could be like a balanced meal on earth.

An astronaut can store enough food for one week in a package about the size of a shoe box. Most foods are freeze-dried

Study the balanced meals with your parents. Plan a meal that you could eat in space.

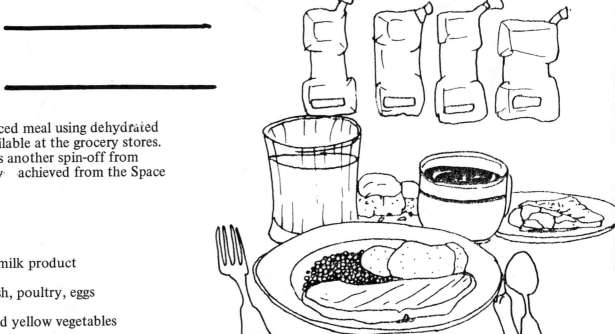

Create a balanced meal using dehydrated baby food available at the grocery stores. This product is another spin-off from the technology achieved from the Space Program.

Child
Serving

3   milk or milk product

1   meat, fish, poultry, eggs

1   green and yellow vegetables

1   citrus fruits and tomatoes

1   potatoes, other fruits and vegetables

3   bread, flour and cereal

Nutrients in 1 or 2 cups of milk daily can be satisfied by cheeses or ice cream
Meat, fish and poultry may be alternated with eggs or cheese, dried peas, beans
3 to 5 cups of fluid daily is recommended.

Help the child learn to group foods into the basic four food groups: Fruit and vegetables
breads and cereals, meat and dairy
Help the child learn to set the table
Help the child learn to follow directions as meals or recipes are followed
Help the child recognize various foods in the grocery store
Help the child learn proper nutritional habits, ie eating basic four foods instead of candy, cokes and junk food.

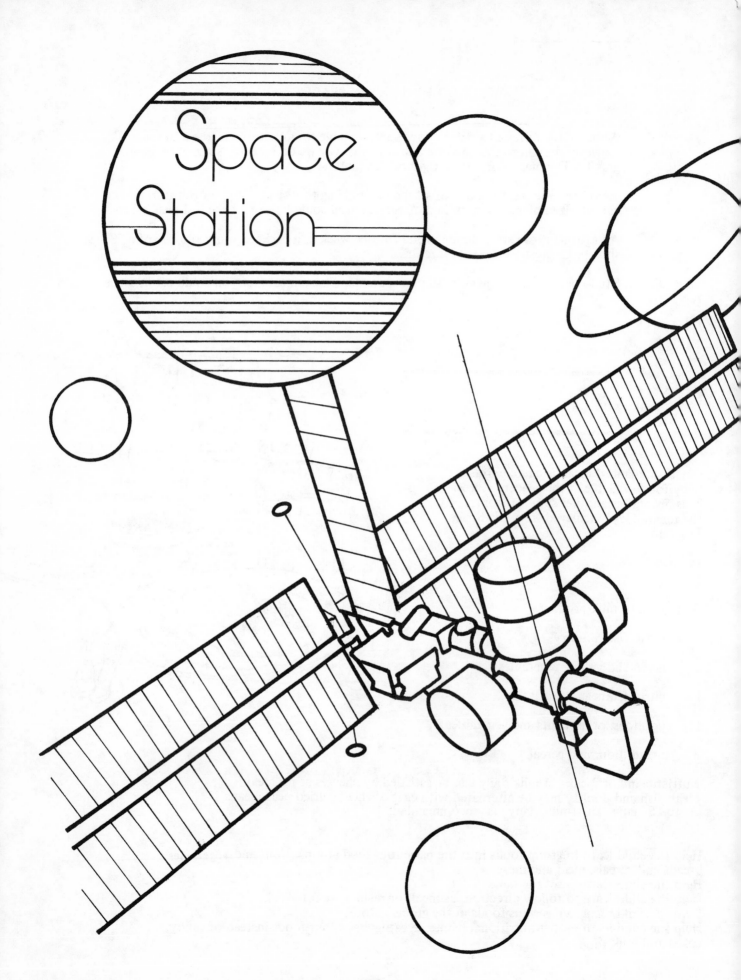

Space
Station

NAME: Space Station

SUBJECT/SENSE: Science (Touch)

SKILL: Dramatize being in space

PROCEDURE:
1. With the children, build a spaceship link up and dramatize a trip into space. A large box can be made into a spaceship or shuttle. Cover with butcher paper, paint, cut out windows, etc. The children can be creative in designing their space ship.
   Materials: box, paper, paint, scissors

2. Encourage the children to talk about an imaginary trip into space, stopping at your space station. Encourage the children to use their imagination above what they would see in space.
   Materials: none

3. One of the experiments Christa McAullife, the first teacher selected to go into space, had planned to perform in space was the Carnation (celery) experiment of osmosis. The teacher will put food coloring in a glass of water then put the Carnation in the water. The children will observe the white Carnation turn the color of the food coloring because of osmosis.
   Materials: white Carnation, (celery stalk) food coloring, glass of water

4. Have the children act out movements described in the record *Captain Entropy*. (Bruce Haack, Dimension 5 Records for Children, Box 185, Kingsbridge Station, Bronx, New York 10463)
   Materials: record, record player

PARENT/CHILD EXPERIENCE:
Your senses are very different in space. Since there is zero gravity, your body becomes disoriented. Try these activities with your child. Sit in a chair and close your eyes. Place your hands over your knees then open your eyes. Ask, how did you know where to place your hands? (He did not use his eyes or sense of sight or touch.)

OBJECTIVE: Participate in the activities

BACKGROUND INFORMATION:
Soon a Space Station will be built in space where the experiments done in the Shuttle can be done all the time, attended by people who could stay in orbit for months at a time. Would you like to live on a Space Station?

SPACE MAIL

NAME: Space Mail

SUBJECT/SENSE: Science (Sight, Hearing)

SKILL: Enhance science vocabulary

PROCEDURE:

1. The teacher can make the word family Space Shuttle to increase the initial sound development of the children. Write any word on the Shuttle omitting the beginning letter (sound). As the strip is pulled slowly through the window new words appear. The children can be encouraged to learn word families as new words appear.
   Materials: folder pattern, scissors, markers

2. The teacher can use various vocabulary words to encourage the children to develop language experience stories. Large sheets of paper can be given to the children to illustrate their space concepts. The teacher can write their stories on the pictures. Then the sheets can be taped together and attached to paper coat hanger cardboard tubes. A window can be cut into a box representing a TV screen. The tubes then slipped through the box creating a TV script. Children will look at their TV show over and over.
   Materials: paper, markers, coat hangers, tubes and a box

PARENT/CHILD EXPERIENCE:
   Encourage your child to develop his vocabulary using space words over and over.

OBJECTIVE: Use new words.

BACKGROUND INFORMATION:
   Word families:

-an	-am	-it	-ill	-et	-ing
man	ham	hit	Bill	bet	sing
can	jam	bit	dill	get	ring
fan	ram	fit	hill	jet	ding
pan	Sam	sit	fill	let	ping
van	dam	pit	sill	met	wing
ran	yam	wit	mill	pet	king

   Sample Vocabulary:

ballon	airline	water	planet	lift-off	airmen
helicopter	wings	weatherman	star	gravity	runway
glider	propeller	cumulus	space suit	thrust	hangar
jet	snow	cirrus	astronaut	zoom	ticket
rocket	rain	stratus	blast-off	airport	fuel
airplane	hail	rocket	countdown	aviation	pilot

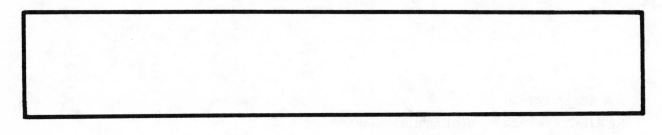

WORD FAMILIES

The strip is slowly pulled through the shuttle allowing various words to be formed.
Various math combinations that total the same sum (or difference) may be substituted.

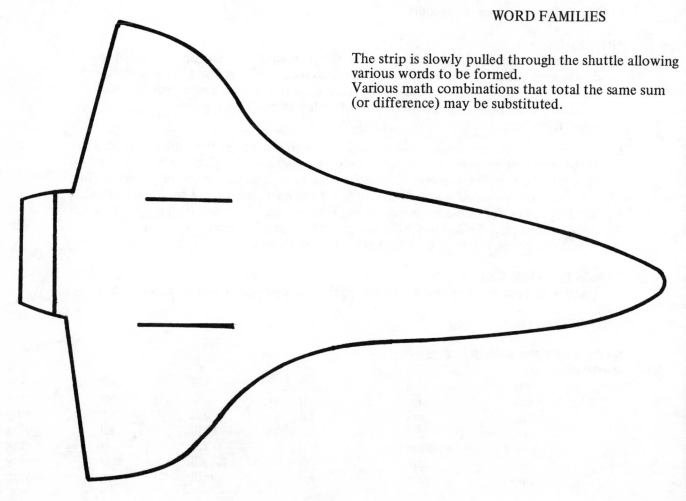

Cut on slits and insert strip. Write your initial letter on the strip and word family on the shuttle

BULLETIN BOARDS

NAME: Bulletin Boards

SUBJECT/SENSE: Science, Math (Sight)

SKILL: Reinforcement of concepts presented

PROCEDURE:
1. The teacher can display the childrens' artwork as she develops bulletin board concepts based on appropriate educational objectives. Various concepts: vocabulary, number concepts, etc. will be reinforced as displayed on the bulletin boards.
Materials: paper, art supplies (See Suggestions)

2. As the children draw their space pictures the teacher may use the ideas to develop bulletin boards or additional folder games.
Materials: paper, drawing instruments

PARENT/CHILD EXPERIENCE:
Encourage your child to draw space pictures and add words to the picture as he tells about the picture.
Materials: paper, markers, crayons

OBJECTIVE: Draw and talk about space pictures

Contributed by Dr. Anne Christenberry, Augusta College

98

NAME: Puzzles

SUBJECT/SENSE: Science (Sight)

SKILL: Enhance fine motor development through puzzles

PROCEDURE:
1. The teacher can laminate pictures from the resources received, then cut into puzzle pieces. Then the children can put the puzzle together.
   Materials: Resource pictures, scissors

2. The teacher can use the following symbols to create a matching game. Laminate one set of symbols onto the file folder. Cut out the other set and allow the children to match the symbols. Younger children can match the symbols easier if they are color coded, i.e., color both the rockets red, both balloons blue, etc.
   Materials: Aerospace symbol patterns, colors, scissors

3. The teacher can glue a set of large numerals to a file folder. Cut another set into pieces. Allow the children to match the numeral pieces. Be sure the numerals are color coded for younger children.
   Materials: file folder, numeral patterns

PARENT/CHILD EXPERIENCE:
   Make puzzles out of your childs favorite pictures. The cereal box covers can be cut into pieces and make ideal puzzles.
   Materials: Cereal box covers, scissors

OBJECTIVE: Complete the puzzles

BACKGROUND INFORMATION:
   Suggestion for puzzle pieces:
   For three-year-olds: 4 puzzle pieces are suggested;
   for four-year-olds: 12 pieces;
   for five-year-olds: more than 12 pieces

Glue one set of aerospace symbols onto the file folder. Cut out the other set and allow the child to match the symbols. Color matching symbols for younger children and they can match the symbols by color and shape.   (Enlargements of the symbols are found under A Trip to the Airport)   p. 55

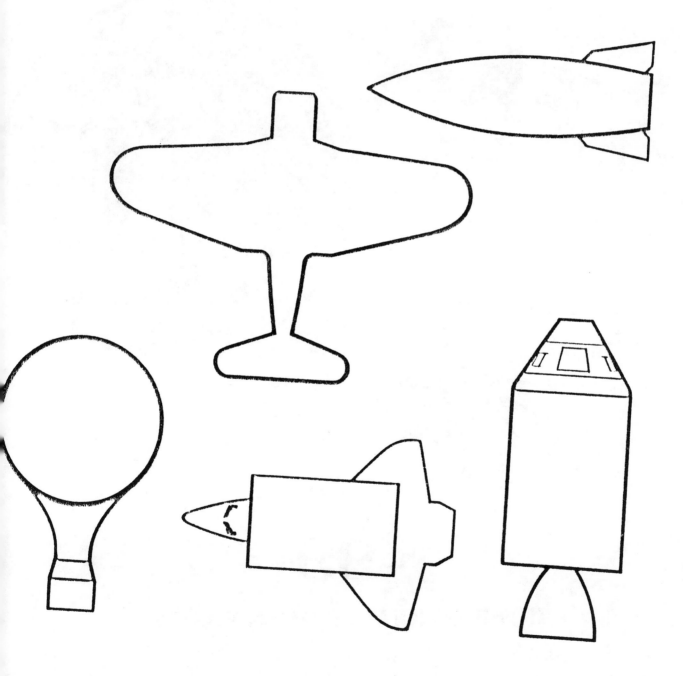

the **99**s

INTERNATIONAL WOMEN PILOTS

NAME: National Aeronautics and Space Administration (NASA)

SKILL: Awareness of the space program

PROCEDURE:
1. Numerous educational materials, charts, posters, and pictures can be obtained from NASA by being listed on the teacher mailing list.   Write to NASA, Educational Publications Services, LEP, Washington, D.C. 20546.

2. Try to visit Kennedy Space Center or another NASA facility. Visit the souvenir shops and teacher resource room to obtain educational material.

3. Contact NASA to find out about the possibility of the Space Mobile visiting your school.

PARENT/CHILD EXPERIENCE:
Watch the newspaper or news programs on TV that relate to space and discuss the happenings with your child, as appropriate.

OBJECTIVE: Watch for current events

BACKGROUND INFORMATION:
After the Soviet Union launched the first artificial satellite, Sputnik, the Space Age was born. Military programs in the 1950s developed rockets for our new space program. Then the National Aeronautics and Space Administration (NASA) was established to plan to manage the new Space Program. NASA supports educational programs for teachers by providing materials of instruction, aerospace workshops, teacher resource rooms located throughout the U.S. These resource rooms contain over 5,000 slides, video tapes, and print material. Contact the address above for the nearest teacher resource room to you.

NAME: Federal Aviation Administration (FAA)

SKILL: Awareness of aviation careers, educational materials

PROCEDURE:
Integrate aviation into the classroom through the basic skill subjects. Obtain free education materials by writing:

FAA, 800 Independence Avenue, S.W., Washington, D.C. 20591

PARENT/CHILD EXPERIENCE:
Request the order form from the above address. With your older children talk about various careers in aviation and how important it is to study the basic skills so they can have an exciting career in aviation/space.

OBJECTIVE: Learn about careers in aviation/space

BACKGROUND INFORMATION:
In 1966 a law was passed to create the Department of Transportation and the Federal Aviation Agency (FAA) became the Federal Aviation Administration. One of the principle functions of FAA is air traffic control which involves almost half of all its personnel. The continuing mission of the FAA is to provide the safest, most efficient air transportation system in the world. The FAA provides aviation education materials and support. It is organized into eight regions. Contact the address above for your region director of aviation education.

NAME: Civil Air Patrol (CAP)

SKILL: Awareness of the Cadet Program teacher resources, and support services

PROCEDURE:

1. Order the educational materials from CAP. Write: HQ CAP-USAF/PA, Maxwell AFB, AL 36112

2. Request information about the National Congress on Aviation and Space held each year for teachers.

3. Contact a local CAP squadron for assistance in speakers or field trips.

PARENT/CHILD EXPERIENCE:
Learn about the cadet program for children 13 through 18 years of age. The squadron usually meets once a week and can provide numerous opportunities for your child. Contact the above address for the squadron closest to you.

OBJECTIVE: Learn about the Cadet Program

BACKGROUND INFORMATION:
The Civil Air Patrol was established in 1941, but in 1948, through passage of Public Law 557, the Civil Air Patrol became the official auxiliary of the United States Air Force. This act enabled CAP to receive Department of Defense excess property and Air Force assistance. The CAP service the community and country through its three primary missions: emergency service (which are search and rescue, natural disasters and civil defense), aerospace education and cadet program. CAP is organized into eight regions, which include all 50 states, Puerto Rico, Virgin Islands, and District of Columbia. Each region has a Director of Aerospace Education that will assist the classroom teacher in providing aerospace education materials, support and field trips. Contact HQ CAP for your region director.

NAME: The Ninety-Nines, Inc.

SKILL: Awareness of the International Women Pilots Association

PROCEDURE:
Write The Ninety-Nines, Inc., International Women Pilots, Will Rogers World Airport, P.O. Box 59965, Oklahoma City, OK 73159, for information and catalogue of available resources (including this text)

PARENT/CHILD EXPERIENCE:
Obtain information on women pilots and discuss aviation career opportunities with your (female) child.

OBJECTIVE: Career awareness

BACKGROUND INFORMATION:
The Ninety-Nines can trace the history of their organization to 1929, although women in aviation can be traced many years before that. As early as 1784, Madame Elizabeth Thible ascended as a balloon passenger in Lyons, France. Women were piloting balloons by 1789. Raymonde de Laroche was only 23 in 1909 when she became the first licensed female pilot (France) and then Harriet Tuemby became the first American woman to receive her license in 1911. In 1929, there were more than 100 licensed women pilots and all 117 licensed women pilots were sent organizational letters. When the charter deadline ended, there were 99 charter members, thus the name Ninety-Nines. Amelia Earhart became the first elected president in 1931. They formed the first Woman's Air Derby and this annual event known as the Powder Puff Derby, ran until 1977. Ninety-Nines sponsor the U.S. Precision Flight Team and present the Amelia Earhart Memorial Scholarship Award and the Amelia Earhart Research Scholar Grant.

Duplicate the following parent activity pages. (Make enough copies for each parent). Cut out each activity and send home with the children as you present the concept in class. (Remember, you may choose to present one or more of the activities or none, depending on your interests and the level of your children.)

NAME: Clouds

PARENT/CHILD EXPERIENCE:
Go outside and look at the clouds with your child. Ask your child what he thinks the clouds mean. If some clouds look darker than others ask your child if he knows why some clouds look darker than others. If there is a dark or thicker cloud (it is more dense allowing less sunlight to pass through) it probably means it is going to rain because it has more moisture in it. Ask your child to talk about how the rain sounds.

OBJECTIVE: Observe various clouds

NAME: Weather

PARENT/CHILD EXPERIENCE:
Hot air expands. This can be demonstrated by the following experiment. Blow up a balloon. Tie it with a string. Tie it to a hot radiator or near a sunny window. When the balloon gets hot, it bursts because hot air expands. Ask your child to tell what happened. Materials: Balloon, string

OBJECTIVE: Observe hot air expanding

NAME: Constellations

PARENT/CHILD EXPERIENCE:
Go outside and observe the constellations with your child on a clear night. See if you can locate the Big Dipper.

OBJECTIVE: Observe stars in the sky

NAME: Solar System

PARENT/CHILD EXPERIENCE:
Go outside with your child at different times during the day. Watch the effects of the sun casting a shadow at different times of the day. This will show your child how the sun changes the amount of heat and light provided during the day.

OBJECTIVE: Show that the sun gives us heat and light as it shines on earth.

NAME: Kites

PARENT/CHILD EXPERIENCE:
1. The parent will read the safety rules of kite flying to their child.

2. Build and fly a kite with your child.
    Be sure to attach a tail to the kite. It keeps the bottom of the kite down providing stability. If you don't have a tail, the kite spins. The length of the tail can be adjusted to provide the stability needed.

OBJECTIVE: Fly a kite seeing the effects of air

NAME: Balloons

PARENT/CHILD EXPERIENCE:
1. Have your child blow on a dandelion (spring or fall). Have your child watch the flight of the dandelion and compare it to balloons and birds in flight.
    Materials: dandelion, balloon

2. Help your child fill one balloon with air and another with water and tie the balloons. Have your child take the two ballons outside and drop them at the same time to compare which is heavier; lighter. Explain the concept of gravity to your child — heavier objects fall faster on earth.

OBJECTIVE: See the effects of air and gravity

NAME: Parachute

PARENT/CHILD EXPERIENCE:
Find different scraps of cloth at home. With your child make parachutes out of different materials and add different weights. Have your child compare the difference as the parachutes fall. (The cloth will be too heavy unless a heavy weight is used.)
Materials: Klennex, scraps of cloth, string, weights

OBJECTIVE: Be able to drop the parachute seeing the effects of air on a falling object

NAME: Flight

PARENT/CHILD EXPERIENCE:
Help your child draw an airplane, noting the front (that thrust makes it move forward) and top (that lift makes it go upward). Talk about how a plane flys. (An airplane flies because air lifts it up, gravity pulls it down and the propeller and engine pull it forward.)
Materials: Paper, crayons

OBJECTIVE: Draw an airplane

NAME: Pilots

PARENT/CHILD EXPERIENCE:
1. Try to visit a local airport with your child. If a pilot is available he may show you the parts of an airplane. Your child can compare the different airplanes at the airport.

2. After the airplane lands it may to to an airport garage called a hangar. The child can make airplane hangars out of boxes and count small model airplanes as they enter into their hangars. If boxes and model airplanes are not available the folder game pattern can be used to create this game. Numeration to ten will be enhanced.
Materials: Boxes, model airplanes or folder game patterns, scissors, markers

OBJECTIVE: Visit an airport; place airplanes in hangars

NAME: Airport jobs

PARENT/CHILD EXPERIENCE:
Encourage your child to tell about an imaginary airplane trip. Encourage him to tell about where he is going, what he would find, and why he wants to go there.

OBJECTIVE: an airplane trip

NAME: Aerospace Pioneers

PARENT/CHILD EXPERIENCE:
Talk to your child about aviation when you were his age. How did you travel compared to how people travel today?

EXPECTED OUTCOME: Listen and answer questions

NAME: Rockets

PARENT/CHILD EXPERIENCE:
The parent will blow up a balloon to demonstrate that a rocket needs thrust to blast-off. As your child lets a balloon go the air is released from it causing it to move forward.
Materials: Balloon

OBJECTIVE: Listen to the rocket balloon

NAME: Rocket

PARENT/CHILD EXPERIENCE:
Help your child made a rocket out of a cardboard cylinder (see next page for directions).
Materials: Cardboard cylinder, paper, scissors, glue or tape

OBJECTIVE: Create a rocket

NAME: Collage

PARENT/CHILD EXPERIENCE:
The parent will help your child cut out magazine pictures or draw pictures of things that fly. Using a variety of media (colors, paint, drawings, pictures, etc.) encourage your child to make the collage as creative as possible. The child's concept of things that fly will be enhanced.
Materials: Magazine, paper, scissors, markers, available drawing tools

OBJECTIVE: Crate a collage

NAME: Trip to the Moon; Blast-Off

PARENT/CHILD EXPERIENCE:
Parents can observe the moon phases at night with their child. Talk about how the moon changes shapes. Encourage your child to draw these shapes.

OBJECTIVE: Observe the moon

NAME: Astronauts

PARENT/CHILD EXPERIENCE:
With your child make a moon landscape. Place small objects on a piece of cardboard or other hard surface. Cover with tin foil. This will create a surface of craters and mountains. Purchase 10 small toy astronauts, space ships, etc. and encourage your child to act out the moon landings.
Materials: Tin foil, cardboard, small objects, toy astronauts

OBJECTIVE: Create a surface representing craters and mountains

NAME: Space Sensations

PARENT/CHILD EXPERIENCE:
The astronauts see a sunrise and sunset every 45 minutes as they orbit the earth. An astronaut day is the same as a day on earth — there is still 24 hours in a day, however, we see the sunrise and sunset at different times. With your child, watch the sunrise and sunset. The child will learn the day (sun) and night (dark) concept.

OBJECTIVE: Observe a sunset and/or sunrise

NAME: Space Shuttle

PARENT/CHILD EXPERIENCE:
The parent can set up a tent at home and let your child participate in various activities — eating, paper work (coloring, etc.), playing, etc. Your child will learn about living in a confined space. Compare this to living in a confined space. Compare this to living and working in Space Shuttle.
Materials: Bed sheet

OBJECTIVE: Create a tent and become aware of limited space

NAME: Space Suits

PARENT/CHILD EXPERIENCE:
Talk with your child about how layers of clothing insulate the body to keep you warm. The astronauts traveling on the moon needed insulation because there was not atmosphere and they needed to keep their body temperature constant so they would not die. (They also needed an air supply since there was no oxygen to breath.) Let your child dress up as an astronaut, encouraging him to use his imagaination.

OBJECTIVE: Dress up as an astronaut for dramatic play

NAME: Space Food

PARENT/CHILD EXPERIENCE:
Have your child help prepare balanced meals at home using dehydrated baby food. Talk about the 4 basic food groups. (See parent note)
Materials: Dehydrated baby food (Hines or Gerber)

OBJECTIVE: Prepare then compare taste, smell, texture of food

NAME: Space Station

PARENT/CHILD EXPERIENCE:
Your senses are very different in space. Since there is zero gravity, your body becomes disoriented. Try these activities with your child. a. Sit in a chair and close your eyes. Place your hands over your knees then open your eyes. Ask, how did you know where to place your hands? (He did not use his eyes or sense of sight or touch.) b. Close your eyes again. Put your fingers in your ears. Again, question your child to see how he knew where his ears were.

OBJECTIVE: Participate in the activities

NAME: Space Mail

PARENT/CHILD EXPERIENCE:
Encourage your child to develop his vocabulary using space words over and over.

OBJECTIVE: Use new words   108